Down to Earth

Celebrating a Blessed Life on the Land

Curt Arens

PUBLICATIONS

Down to Earth
Celebrating a Blessed Life on the Land
by Curt Arens

Edited by Gregory F. Augustine Pierce
Cover design by Tom A. Wright
Text design and typesetting by Desktop Edit Shop, Inc.

Scripture quotations are from *New Revised Standard Version Bible: Catholic Edition*, copyright © 1989 by the Division of Christian Education of the National Council of the Churches of Christ in the United States. Used with permission. All rights reserved.

The "Prayer in Honor of St. Isidore" and "The Litany of St. Isidore" are from *Novena in Honor of St. Isidore: Patron of Farmers*, copyright © National Catholic Rural Life Conference, 1954. The quotations from "Catholic Social Teaching and Agriculture" are from *For I Was Hungry and You Gave Me Food: Catholic Reflections on Food, Farmers, and Farmworkers*, copyright © United States Conference of Catholic Bishops, 2004. Used with permission. All rights reserved.

Published by ACTA Publications, 5559 W. Howard Street, Skokie, IL 60077-2621, (800) 397-2282, www.actapublications.com

Library of Congress Catalog number: 2007942128
ISBN: 978-0-87946-347-3
Printed in the United States of America by Versa Press
Year 15 14 13 12 11 10 09 08
Printing 8 7 6 5 4 3 2 First

Contents

Dedication

To my wife, Donna,
our children, Lauren, Taylor and Zachary
and my parents, Harold and Margaret.

A Note from the Publisher

I couldn't tell you the difference between a combine and a thrasher. I don't think I've ever spent a night on a farm, although my mother grew up on one and so did my wife's father. But we're all city folk now, because that's the way the world has "developed" and continues to "develop."

Still, there is all that land out there—between cities, I mean—that has acres upon acres of wheat and corn and livestock and barns. And every once in a while you see an actual farmer out in the field and you say to yourself, "That's right, somebody has to grow all that food we eat." And maybe you even say to yourself, "I wonder who these farmers are and what they think about?"

What I am especially interested in is what they might think about religion and spirituality. So when Curt Arens proposed his book, *Down to Earth*, as part of ACTA Publications' ongoing commitment to publishing first-person reflections by ordinary laypeople in what we call our American Catholic Experience series, I was delighted. For Curt Arens is Catholic in the same way he is a farmer: it is in his blood; he drank it with his mother's milk; it is who he is at the deepest level of his identity.

Those of you who are looking for a political polemic on the issues facing modern agriculture or a theological treatise on Church doctrines or practices are going to have to look elsewhere. Curt Arens is just going to tell his story. It centers around the farm that lies along Bow Creek in northeastern Nebraska, which has been in his family for four generations, and the Catholic churches he and his extended family have attended and supported for decades.

Arens narrates how he met his wife, Donna, at a fundraiser for the Catholic school where she taught, and how they are raising three children, have helped care for aging relatives, and run his family's farm together. He describes attending "Dirt Campus" at the University of Nebraska and taking over the farm before he was ready and trying to make ends meet even today. And he shares how he learned to pray and made Cursillo and had his kids baptized and buried his loved ones, all in his beloved Catholic church.

Down to Earth is about the harvest of blessings that come from trying to help bring about the kingdom of God on earth, as it is in heaven. In that sense, it is for all of us, no matter where or how we find ourselves living and working in this world. It is about every farmer, and it is about you and about me.

Sit back and enjoy it.

Gregory F. Augustine Pierce
ACTA Publications

Introduction

As I gaze across our farmyard on a still evening at a sky filled with stars, awe at the dark expanses of space makes me feel that I am but a tiny speck in a magnificent universe. On this farm where I live and work, my grandfather probably looked up from the steps of the old farmhouse and felt the same way. In those days, farmers worked directly under that grand universal sky every day, walking behind horses plowing sod under the burning sun or strolling on a green mat of grass tending cattle on the prairie. Nature wasn't something they had to look for. Farm animals, wildlife and the elements of wind, sun, rain and snow and all of God's creation were right there before them.

Today, we farmers are often confined behind steel and glass, operating tractors from a pilot's seat in an air-conditioned, stereo-filled cab, gliding above the soil. There are days when a farmer might leave the breakfast table, walk a few feet to a tractor cab and work from that cab, driving back and forth, up and down row upon row of crops all day, without stepping foot on the soil or breathing air in the open. The farmer might spend the entire day behind a desk or at a computer monitor, checking grain prices, updating farm records and feed rations or taking inventory of livestock.

Rural lifestyles are viewed by urban dwellers as idyllic, but the truth is not always as clear or simple. Farming is one of the most difficult occupations there is. Many farms in northeastern Nebraska where we live, which operate on 600 acres as we do, are too big to be run by hobby or part-time farmers but too small to provide a family living solely from the land. Many of us family farmers send our spouses off to work in town, our young children to a baby sitter or to school, and work another job ourselves—sometimes just so we can obtain health insurance for our families. I look around my community knowing that many of the farms are unoccupied during the day because the farmer and his family are all working elsewhere and trying to farm—a full-time job in itself—in their "spare time." Rising costs and volatile farm commodity prices have changed the economics of family farming in recent years. To make a living solely from a farm with conventional production methods, farmers have to cover four times the number of acres my parents did when they farmed here.

> *For most of us, the blessings of a life on the land far outweigh the challenges we face.*

For most of us though, the blessings of a life on the land far outweigh the challenges we face. My wife, Donna, and I feel honored to carry on a family tradition as the fourth generation of Arenses to work our farm. Donna teaches English, Reading and Literature and is the eighth-grade homeroom teacher at St. Rose of Lima School in our hometown of Crofton. It's the same school I attended as a youngster. In addition to farming full time—raising cattle, hogs, corn, soybeans, oats, wheat, alfalfa and a few Christmas trees—I work as a freelance agricultural journalist to help cover the bills. Our children, the reason why farm life is so important to us, are quite young. Daughter Lauren is eight years old and a second-grader at St. Rose School. Daughter Taylor is six and has just started school. Son Zachary is two and already owns a stable of toy tractors, just as his daddy did when he was young.

Farmers today may not be as close to nature and the Almighty Creator as farmers of the past, but we are still more in touch with the ecosystems God created than most people. When Pope John Paul II was introduced to a farmer during his visit to the Living History Farms near Des Moines, Iowa, in October of 1979, he told him, "We are all farmers." Of course, the great pontiff was speaking of sowing the seeds of faith. He compared his vocation of fulfilling the legacy of Saint Peter with the farmer's legacy of sowing seeds and producing food, all with God's abundant help. He knew what he was talking about, because long before Karol Wojtyla was elected pope his first priestly assignment was as an assistant pastor at the rural Assumption of Our Lady Church in the isolated Polish village of Niegowic. While serving there, he not only celebrated daily Mass but lived among the farm folks from his tiny parish, taught religion in the region's five elementary schools, worked beside farmers in the fields, and made regular pilgrimages into the countryside to pray. Pope John Paul II knew there was something inherently sacred about rural life.

I can say with accuracy that I am an American farmer, a family farmer, and a corn farmer. But because I truly believe that my faith and my farming are all of one piece, I can also say that I am a Catholic farmer. I used to think my Catholicism and my work were separate things, but after years of farming, I now understand my spiritual life and my occupation are molded together in a way that cannot be separated. I haven't come to this conclusion through some single epiphany. The people of faith around me—my wife and children, my parents and in-laws, my extended family, and of course our parish priests and sisters as well as our neighbors, close friends and fellow parishioners—have provided me with hundreds of experiences proving to me that my vocation, my calling in life, is to work and raise a family on the land.

In my mind, the real connection for Catholic farmers between God and our work and family life comes in the Eucharist. Of all the mediums Jesus could have chosen in which to be truly present with us, he chose bread and wine—"fruit of the vine and work of human hands" as we say in the Offertory of the Mass. The fact that he selected these simple products of the land not as symbols but as true mediums for his body and blood solidifies my devotion as a Catholic and as a farmer. I see the Eucharist as a sign that those of us caring for the land, raising crops, and tending livestock, are truly blessed in a special way.

Sharing my faith stories in this book has been a difficult undertaking for me, because I have always viewed faith as a private matter. I have written this book as an imperfect man, a farmer who has a few weedy fields of my own (just ask my neighbors), and a husband, father and lay person still seeking answers to my faith questions. But I hope that when you read these stories you will observe—on my farm and in my fields, with my family and in my rural faith community—the true blessings of creation that I cherish so deeply.

Curt Arens
The Arens' family farm
Crofton, Nebraska

Don't Sweat the Details

The chilly, northerly breeze was bracing as I stepped outside. It was an early morning in May with dark clouds hovering. One of my best college buddies, Randy Micek, helped me finish chores around the farm, including bottle-feeding a small orphan calf and checking on hogs in their pens. We trudged up the big pasture hill north of the farm, checking cows with baby calves as we went. We walked silently among the sleeping herd, watching for any newborn calves that might need attention. As we reached the crest of a perfect, grassy vista rising about seventy feet above the peaceful valley that I call home, I paused for a moment, as I might do any other day, and enjoyed a sweeping view of the farm and West Bow Creek meandering below. On this particular morning, I appreciated the quiet, the cows and the farm even more than usual. I prayed silently, asking the good Lord to calm my fragile nerves, because I could feel my heart racing. It was not just any day. It was my wedding day.

"Well, after today, everything will be different," I said, only partly for the benefit of Randy, who was not only a chore-hand-for-a-day but in a few hours would also be one of my groomsmen. I was trying to comprehend what having a bride to bring back home might mean to a bachelor farmer in his early thirties like me.

"You've got that right, buddy," Randy replied with a grin. He understood what I was saying. He appreciated the romanticism of farming, and how much it meant to me to finally have a wife to share that dream with. The rest of the wedding party had stayed the night in Madison, Nebraska, where Donna, my fiancée, was living and where the wedding would occur. But I wanted to spend the night before my wedding at home on the farm, where I have always been the most comfortable. Donna had sent Randy along to assist with chores (and to make sure I'd get to the church on time for photos, I'm sure). I had no apprehension about marrying Donna. I was not more certain of anything in my life.

On my wedding day, I was concerned not only about how the threatening rain clouds might affect our wedding plans but also how they might affect our livelihood.

I worried, however, about leaving the cow herd to honeymoon for a few days, although I knew things would be in good hands with my Dad on duty. My greatest anxiety that morning was centered on something all farmers are obsessed with – weather. It can make or break a farmer's livelihood. Rain (or the lack of rain), hail, wind, snow and mud all affect the operations of a farm and the bounty of harvest at the end of the growing season. On my wedding day, I was concerned not only about how the threatening rain clouds might affect our wedding plans but also how they might affect our livelihood.

I comforted myself by recalling the words of Father Ron, my parish priest at St. Rose at the time. "Don't sweat the details," he told Donna and me when we excitedly announced our engagement a year earlier. "Worry more about the marriage than the wedding day. All you need to get married are the two of you, a priest, and a couple of witnesses." His words put my mind at ease a bit, but I was still anxious to get to Madison. Randy and I walked down the hill. Cows were quietly munching grass, with small calves sleeping peacefully beside their mothers. All was well, as it should be. As God wants it to be.

I had met Donna while golfing in a best-ball tournament in Madison with college friends who lived there. I didn't often make the trip from my home near Crofton south down U.S. Highway 81 to Madison. However, Kurt Jackson, a buddy who grew up just over the hill from our farm, and his wife, Lisa, lived in Madison and occasionally invited me to visit. The golf tournament was a fundraiser for St. Leonard Catholic elementary school where Lisa taught. Some of the other teachers at St. Leonard, including my future wife, sold sandwiches to golfers as we passed along the tournament route. I met Donna, wondering immediately to myself if she was married. I didn't see a ring, and that was encouraging, but I knew I couldn't impress her with my stunning golfing ability, because I had none. Although our encounter was brief, I noticed her quick smile. Kurt evidently noticed me noticing Donna.

A month later, Kurt and Lisa invited me for supper. When I arrived a little early, Kurt sat down and flashed a grin like the cat that had just swallowed the canary. "There are going to be a few other people here too," he told me. By "other" he meant Donna and another teacher from St. Leonard whom she brought along for moral support. I was too shy to speak much to Donna that evening, but I was charmed by her bright, blue eyes and quick sense of humor.

It was harvest time, and I was busy in the fields until late every night. I planned to call her the following week, but after three days Donna called me. "Weren't you ever going to call me?" she asked, sounding a little perturbed.

"Sure," I replied. "But during harvest, there isn't time for much else."

I don't think Donna believed me, but she learned quickly after we were married that fieldwork consumes a farmer during both planting and harvesting and that missing even several hours' work at those times can often cost thousands of dollars in crop loss due to bad weather later. But I gave up a night of harvest a few days later, taking Donna on our first date. We had supper at a little Mexican place in south Norfolk, Nebraska, and went to a movie called *The Chamber.* (Leave it to me to pick a date movie about a former Ku Klux Klansmen sitting on death row.) Donna's spirit wasn't dimmed by the gloomy story line, however. Driving around Norfolk after the movie, we started sharing deep thoughts and beliefs, as if we'd been friends for years. It was unusual for an introvert like me to feel comfortable around someone so quickly. Our talk naturally turned to our mutual faith, because being Catholic was—and still is—such an important part of both of our identities.

I kept thinking that Donna was prying things out of me that I don't normally share with anyone. Usually I was most comfortable in the cab of a combine, with only rows of corn ahead of me to think about. What was happening while we drove around town was outside my comfort zone. I realized I was sharing my beliefs willingly with her about marriage, about the sanctity of life, about my hopes for a family of my own, and about the family farm that is so dear to me. I was amazed at her frankness and was relieved that we stood on the same ground with all of those important questions.

Donna grew up in Omaha, which for a farm boy like myself is considered "the big city." Her mother had always told her that she should marry a farmer because, as a kid, she always loved gardening and animals. While Donna might have had a hidden desire for life on the land, however, she had no practical knowledge of a working farm. She was anxious though, to take it all in. We spent hours riding together in the tractor and combine, building our relationship as she learned the work of farming. Our courtship was also built on faith. We enjoyed attending weekend Mass together and went to several St. Leonard parish activities, including a number of Lenten fish-fry nights at the Knights of Columbus Hall.

One night about five months after our first date, we were shopping at the mall in Norfolk. We started talking about engagement rings.

"Well, let's stop talking about this and just go over and buy them," I said. Donna's jaw hit the floor. "You're kidding, right?"

"Nope," came my reply. We had been discussing marriage since our very first date, and I was old enough to know what I wanted. We walked into the jewelry store and bought Donna's engagement ring and our wedding bands. I didn't have the money to pay for them that moment, so I put down a first payment on the rings, telling the clerk that I'd be back when I had the rest of the money. I told Donna that it could be months before I could afford the final payment.

Donna agreed to become my wife, although neither of us had any real idea what being husband and wife on a farm would mean.

Still, the deal was sealed. We drove to a park along the Elkhorn River south of Norfolk and stopped behind two gigantic spruce trees. Just as I began to officially pop the question, a city police officer drove through the park and shined his flashlight into our car. How I would have hated to be listed in the police report on the night I proposed to my girlfriend! I guess he thought we were innocent enough, so he drove on…and I finished my proposal. Donna agreed to become my wife, although neither of us had any real idea what being husband and wife on a farm would mean.

A few days later, like my father, who sold a load of oats to pay for my mother's wedding ring, I hauled a truckload of soybeans to the local grain elevator and used the money to pay for the rings, without Donna finding out. Then on a cold, late-winter day when Donna was visiting at the farm, I asked her to jump in the pickup to help me check cows for any newborn calves. "They aren't calving yet, are they?" she asked.

"You just never know," I said with a grin, and we drove to the pasture. She looked cross at me, knowing full well the cows weren't even thinking about calving yet. Parking near the stately, century-old cottonwood trees my great-grandfather and his team of workhorses once took shade under while working the fields, I asked Donna to peek in the glove compartment. She opened the glove box door and a little white box popped out into her hand. Donna opened the ring box and gazed a while at the rings we had selected. She couldn't figure out how I had come up with the money to pay for them but was glad she did not have to wait months to get them.

While we were dating, Donna thought those big numbers in my checking account were income. After our wedding, when she began writing checks from the farm account, she realized that those numbers represented debt. Money was one of the very few things we didn't talk much about during our courtship. Donna thought that farmers liked to talk about being poor but with all of the land they owned or worked were actually quite wealthy. That cloudy understanding of family farm finances came to the surface when we spent one Sunday afternoon at Immaculata Convent in Norfolk, completing our required Facilitating Open Couple Communication Understanding and Study Survey (FOCCUS) inventory, a pre-marital quiz for couples planning to marry in the Catholic Church. We were a little nervous about the idea of taking a church-imposed inventory of our personal lives.

We asked ourselves, "What if we don't answer the questions the way they think we should?" We wondered if Donna's parish priest, Father Bill from St. Leonard, might refuse to marry us if we seemed incompatible because of our survey answers. After taking the survey, we drove to the park in south Norfolk and talked about our answers. Through the process, we learned we had answered nearly all of the questions identically—except the questions that dealt with finances!

When we met with Father Bill days later to begin planning our wedding, he hardly mentioned the test, other than to say that we had answered most of the questions nearly the same. He did ask us, however, to talk with each other about finances, particularly about farm and family money issues that might arise later. Living and working in a rural community like Madison, Father Bill probably knew all about the challenges of financially providing for a family on the farm. Donna and I did talk about it at great length, but it was difficult to prepare for all of the unforeseen obstacles that might pop up. However, if we hadn't taken the test, we probably wouldn't have realized the importance of that issue.

Father Bill met with us about the wedding liturgy, but most of our marriage preparation was done by Fred Ridder, deacon at St. Leonard, and his wife, Millie. They were both farmers, so we hit it off right away. Over the several weeks of sessions we had at their farm home, we grew to truly respect them not only as friends but also as a married couple. The advice of Father Ron rang through as the Ridders shared their experiences, not only relating to finances but also about raising children, keeping romance in the marriage, how to disagree fairly, and—important to us—about faith.

After finishing our chores, Randy and I jumped in my pickup and drove to Madison, arriving at St. Leonard's in plenty of time for wedding photos. I joked with my groomsmen that I should have worn my farmer's pliers and holster—as I would on any typical day on the farm—under my tuxedo coat, just in case we needed to fix some candleholders or floral arrangements with baling wire. For a farmer, a pliers is like another right arm. Donna scoffed at my comments, but fortunately my brother, Paul, who was my best man, had brought his pliers, because just before Mass began, he and Connie Dentlinger, our matron of honor, were able to fix a bouquet of flowers on the altar with Paul's pliers and some wire they found in the sacristy.

No couple could ask for a more beautiful setting for their wedding Mass. St. Leonard is a large, ornate, old church with a statue-adorned altar and an arching cathedral ceiling. The church quickly filled with family and friends. My parents escorted me to the front and I joined my groomsmen near the sanctuary. As the organist began to play and Donna walked down the aisle with her parents, I felt almost out of place. It was like watching a film, where you are a spectator, not part of the cast. Donna took my arm and we walked together to kneel before the altar. A warm smile from Father Bill and a comforting grin from Fred brought the realization to both of us: We were being married and would be husband and wife "till death do us part."

> *I grasped Donna's hand tightly and prayed silently that God would always be one of the cords of this marriage, and that faith might be the cornerstone of the family Donna and I were forming.*

Because we had grown so close to Fred and Millie during our marriage preparation, Fred gave the homily for the Mass. He joked about Donna marrying a farmer and the challenges she might face. He became solemn, however, when he talked about the importance of faith in marriage. The familiar Ecclesiastes 4:12 verse, "A three-ply cord is not easily broken," came to my mind. I looked up and gazed long at the statue of Jesus hanging on the cross before us. I grasped Donna's hand tightly and prayed silently that God would always be one of the cords of this marriage and that faith might be the cornerstone of the family Donna and I were forming.

After Mass, friends surprised us with a horse-drawn buggy ride to the Knights of Columbus Hall where our reception and dance were held. While most folks opt for a ride in a fancy limousine, for Donna and me the simple buggy ride was more fitting. The buggy driver even had a blanket to cover Donna's shoulders as we rode in the open air on a blustery afternoon.

At the wedding dance, Donna's first- and second-grade students from St. Leonard School gathered around her, offering hugs and congratulations. A few of my groomsmen stopped the dance at one point and, to her surprise, sat Donna in a chair in the middle of the dance floor. The crowd exploded with laughter when Kurt and Randy teased Donna about being a farm wife, offering her a pliers and holster of her very own, a five gallon bucket and scoop shovel so she could help feed livestock, and worn out coveralls so she would stay warm during chore time.

When we finally left the hall for our motel room in Norfolk late that night, we were both exhausted. I was worried about who would clean up the dance hall, who would take tables and chairs back to the rental place and what time we needed to be back to Donna's house in Madison the next morning to open gifts. We pulled into the motel parking lot. Still dressed in her mother's beautiful wedding gown, Donna took the key and went ahead to open the motel room door. Still in my black tuxedo, I bent over the box of my pickup to retrieve our luggage.

Up above the parking lot, on a balcony around the upstairs rooms, a bunch of formally dressed teenagers were laughing, evidently partying on their prom night. A couple of girls yelled down at me, "Hey."

I thought they were talking to someone else. "Hey," they yelled again and I looked up. "Did you just get married?" one of the girls asked.

I felt self-conscious in my tuxedo coat with a flower still pinned in place and about the new Black Hills gold wedding band now on my ring finger. I figured they were just making fun of me, but I looked back up at them, a little embarrassed, and said, "Yes, I did."

"That's cool," one girl replied.

It finally sank in that Donna, the person I had been searching for and praying about for so long, was now a complete part of my life.

Pretty cool, indeed.

Bow Creek Chronicles

Joseph Arens was only eight years old in 1867 when his family left the beautiful countryside of Westphalia, Germany, and immigrated to the United States, eventually establishing a farmstead on the wide-open Nebraska prairie in Cedar County, near a stream called Bow Creek. That was my family's beginning on this continent, and it has always been Bow Creek that has tied the family, our faith congregations and our farms together, as Arens family members settled on sections of prairie along this same waterway.

The Arens family and a number of other Catholic families from the same region in Germany founded Sts. Peter and Paul parish, where they built a cathedral on the plains and a little village called Bow Valley. At the time, this was a community where parishioners standing outside of church and visiting after Mass were just as likely to speak German to each other as English.

German Catholic families who crossed the ocean together often settled in their own new villages and parishes, like Bow Valley. Some left Germany because they wanted to avoid their compulsory stint in the German army. Others saw no opportunity in farming the small one- to five-acre ancestral tracts of land they might inherit.

Many of the Catholic families left, however, because the German government at the time assumed a policy that strangled the traditional rights of the Catholic Church. Religious persecution in their homeland seemed to be a catalyst for the new German communities springing up in America, with community life centered around new Catholic parishes and new parochial schools being built upon the ideals of religious freedom. That experience of my ancestors probably has much to do with how deeply farmers in our area still value our rural parishes and schools more than a century later. Even during serious financial challenges on the parish level, local parishioners have regularly pulled together, bridging generations to volunteer time, talent and treasure to maintain our local Catholic institutions.

Families from Knox County traveled by horse and buggy, sixteen miles over some treacherous paths, to attend Sunday Mass at St. Boniface parish.

When Joseph Arens grew into manhood and married my great-grandmother, Mary Kleinschmit, their wedding took place in nearby Menominee at St. Boniface, another ornate Catholic church, made of chalkrock from the adjacent Missouri River bluffs. When my great-grandparents settled about fifteen miles west of Bow Valley, a mile across the county line into Knox County at the west end of Bow Creek, they were among the earliest Catholic families to farm in that particular area.

The town of Crofton wasn't yet established, so families from Knox County traveled by horse and buggy, sixteen miles over some treacherous paths, to attend Sunday Mass at St. Boniface parish. Finally, a new parish called St. Joseph formed at Constance, down Bow Creek a few miles east of Joseph and Mary's farm. Their children, including my grandfather, John Arens, attended Mass at St. Joseph and later on went to the Catholic school there.

John Arens grew up and fell in love with Caroline Luft, a neighbor girl who he had known from church and school. When John and Caroline traveled by train to nearby Hartington to obtain their marriage license before their wedding, my grandfather purchased a beautiful rosary as a wedding gift for Grandma, one she had been admiring in a store window. That rosary was intertwined with Caroline's bouquet on their wedding day.

The rosary became an important prayer for my grandparents throughout their married life. Some friends once told me about a fishing trip they had taken with my grandparents years later. Each couple had their own bedroom in the fishing cabins, and as my friends lay awake in bed they could hear the murmuring of my grandparents through the walls in the other room.

"Boy, they must have a lot to talk about," the couple commented upon hearing Grandpa and Grandma speaking in low tones. As they listened more intently, they realized that my grandparents were praying the rosary together.

As John and Caroline stood together at the altar of St. Joseph Church on a cold, breezy January day in 1916, they prepared to pledge their lifelong wedding vows to each other. John's farm dog, Pup, strolled up the aisle toward the couple. Pup was used to following John everywhere around the farmstead, so he apparently didn't see why a little thing like a wedding should change that. He had evidently followed John from the farm and nudged the church door open with his nose. Grandpa teased Grandma years later that Pup was trying to save his master. I don't think Grandpa needed much saving. He and my grandmother were married for fifty-four years and were blessed with eleven children and scores of grandchildren before Grandma died of cancer in 1970.

After my grandparents were married, they moved onto the farm where Donna and I live now, just a mile west up Bow Creek from where John grew up. There was an old house on the place, but no barn. Grandpa had asked his new bride if she would rather have a new house or a new barn. Always a wise farm wife, Grandma said that the current house was good enough but they couldn't operate a farm without a decent barn to house their horses and milk their cows. So the barn was built first, followed by a number of buildings that were built or moved on the place later and remodeled as a granary, chicken house, hog barn, and ice house. Caroline Arens waited thirty years for her new house, and the house was actually quite old by that time, having been built originally as one of the earliest homes in the town of Crofton and then moved onto our farm—lock, stock and barrel, as they say. It was and still is a large, beautiful house that looks as if it was made for this farm.

The huge, old barn the newlyweds had built first was always the center point of the farmstead. It sheltered the horses that Grandpa used every day for nearly every task on the farm. In the late 1920s and early 1930s, my grandparents hosted barn dances in the spacious hayloft. I've often heard my own father tell stories of his childhood, watching neighbors descended only a generation or two from immigrant families, dance up a storm in our barn.

When John and Caroline's older daughters started classes at a country school, just a mile over the hill from the farm, they didn't know how to speak English fluently. Mostly German had been spoken in the home by my grandparents. Their teacher encouraged my grandparents to start teaching their children English. In the wake of World War I, when American Doughboys were fighting the German army, German families in our area gave up their native language, refusing to speak German in their homes again. Many of the cultural traditions of their former homeland were also abandoned willingly as a way of proving their allegiance to their new country. They didn't abandon their ethnic food, however, and they kept their strong allegiance to their Catholic faith, thank God.

When the town of Crofton was founded in 1892, the few Catholic families there who had been worshipping in nearby Constance founded a parish of their own, St. Rose of Lima, the parish my family still calls our home.

When my father, Harold, and his twin brother, Gerald, were born in the old farmhouse on a bitterly cold, late-July day in 1927, the doctor from Crofton had to travel to check on Grandma Caroline the next morning wearing a buffalo skin coat to keep him warm. That was the only time in anyone's recollection that it froze in July in Nebraska. The corn crop was frozen in the fields. After checking over the newborn twins and seeing the other children bustling about, the doctor told Grandpa, "John, I think you've got something started here you don't know how to stop."

I've always believed that to understand a person, you have to understand where they've been.

New babies always elated John and Caroline, but Grandpa confessed years later that he went out behind the barn that day, away from view, and cried. He had two more mouths to feed in his growing family, had just lost that year's corn crop, and wasn't sure how to make ends meet.

I've always believed that to understand a person, you have to understand where they've been. As I've grown older, I've come to appreciate even more the way my ancestors dealt with challenges. Knowing what they went through helps me keep my own daily financial and family struggles into perspective.

During the drought, devastation and Great Depression of the 1930s, Grandpa was able to hold the family farm together, while many others lost everything. He and a neighbor together took on trucking jobs to help pay the bills. They hauled livestock, grain and potatoes to market, leaving for days at a time. My grandmother took care of the family and the older boys helped with the farming while Grandpa was away.

My father complains that he had to walk to school uphill—both ways—as a child. He's not kidding, because the rural schoolhouse, which still stands, is located on the other side of a high hill east of the farm. Although my grandparents believed strongly in Catholic education, it wasn't practical to transport all the children the four miles to town to St. Rose School. They were needed to help on the farm. They made sure, however, that all of their children attended St. Rose for third and fourth grades so they could receive their First Holy Communion and First Confession while attending Catholic school. The rest of the years, the Arens kids went to religious education classes in town every Saturday.

Although summers were bone dry in the 1930s, the winters were snowy, cold and harsh. Each following spring season, roads were atrocious, with mud up to a car's floorboards. Dad remembers Sundays when it was challenging, because of the weather, to get to town for Mass. He recalls a number of other Catholic neighbors traveling to the Arens farm and piling into the box of Grandpa's old truck for a ride to church together. Dad also remembers the entire family piling into the same pew in St. Rose Church Sunday after Sunday. If you happened to accidentally sit in the Arens family pew on a particular Sunday, you probably received an icy stare from my perturbed grandmother.

Faith was often the only thing that carried the farm families of my grandparent's day through the deep struggles of their time. Sunday Mass and praying a nightly rosary were a few of the ways my grandparents deepened their relationship with Christ. Their faith was often the only thing that helped them get out of bed each morning and deal with their constant financial troubles.

As they grew up, one by one, all of my Dad's siblings started farming on their own or married other farmers. Initially, they each settled within thirty miles of Crofton and all of my uncles and aunts became active members of their local parishes.

Dad was drafted into the U.S. Army in 1950 during the Korean War. Instead of being sent to the front lines in Korea, however, Dad was stationed in Germany, another hot spot of potential conflict with the Soviet Union. Serving in Germany allowed Dad the opportunity to find the old ancestral home of the Arens family. It also afforded him a chance to travel to see the Vatican in Rome. One of the most memorable days of his spiritual life came when he and the other Catholic soldiers on furlough had a private audience with Pope Pius XII. After the audience, they attended the canonization Mass of Saint Anthony Mary Claret and witnessed the pope addressing thousands of worshippers.

Over the years, Dad often has related stories of his time in Rome and his visit to St. Peter's Basilica. As a kid, I loved looking at color slides Dad had taken and reading the old Vatican City guidebook he brought home with him.

He returned home after his tour of duty and started farming with Grandpa and Grandma Arens. He was afforded the same gift and the same demands of farming with his father that I have now. I'm sure Dad and Grandpa didn't see eye to eye on every detail, but today Dad recalls those times with happy, humorous memories.

Dad was in his early thirties and had a few serious girlfriends, but things never seemed to work out in the romance department. So my grandparents were shocked when without warning Dad brought Margaret Bickett home to the farm to meet them.

My mother, Margaret, was born in South Dakota, near a country parish and village called Emmett. Her father, Arnold, was born in a log cabin in Kentucky. Grandpa Arnold was drafted as an infantryman, a "doughboy," and served in France during World War I in the bloody Meuse-Argonne campaign that ended the "war to end all wars." He often attributed his safety in the trenches to his mother's prayers.

When he came home, he worked for Henry Ford in a car factory in Michigan before moving to South Dakota, taking a job as a hired hand for his future father-in-law.

On the farm near Emmett, Arnold met my grandmother, Catherine Kribell, a much younger farmer's daughter, and they were married. Apparently, Catholic education was important to my mother's grandparents too. When Grandma Catherine was a girl, her father sold a load of seed oats to the Benedictine Sisters in Yankton, S.D., each spring to pay tuition bills, so his children could attend Mount Marty Catholic High School.

When Mom first met Dad's parents, Grandma Caroline was in the orchard butchering chickens.

My mother grew up on a farm near Howard, S.D., and attended St. Agatha's High School before studying to be a lab technician with the Benedictines at Mount Mary College in Yankton, which is right across the Missouri River from Crofton, and working part-time at Sacred Heart Medical Center nearby.

Dad met Mom on a blind date, not unlike the way Donna and I first came together. Although Margaret Bickett was much younger than Harold Arens, they fell in love quickly and were engaged after a few months. When Mom first met Dad's parents, Grandma Caroline was in the orchard butchering chickens. Mom remembers Grandma went to her beautiful garden and picked a fresh bouquet of flowers for her.

My parents were married in early May of 1960 at St. Agatha's Catholic Church at my mother's home parish in Howard, S.D. Harold and Margaret moved into a trailer house on Grandpa and Grandma Arens' farm, just a few feet from the big farmhouse, and started a new life together. Mom continued to work at the clinic in Yankton, and

Dad hoped that someday he could work out a transition plan for the farm, so he could purchase the home place from Grandpa and Grandma Arens. That chance came when he and Mom finally purchased the farm in 1965, when I was about a year old. Grandpa and Grandma moved to a house in town, and our little family moved into the main farmhouse. That's just how things happen on a family farm.

The 1960s were extremely dry. Fortunately, Dad milked cows, so we had a weekly milk check to pay bills. In 1968, we received only two-thirds of normal precipitation, and the corn seeds that actually found enough moisture to germinate sputtered into the air a couple of feet but didn't produce grain. My parents scrambled to find enough pasture and hay to feed their milk cows, and the hay they had to purchase was extremely high-priced. In the meantime, late in the summer, my brother, Paul, was born. My parents say his birth was the only good thing that happened during the long, dry growing season. They were thankful the good Lord somehow brought them through it with the farm and milk cow herd intact.

In the early fall of that year, we endured several blizzards. Snowfall for that winter tripled the norm. Roads were blocked most of the time, and Dad had to meet Joe Kremer, our milk truck driver, at the end of our long driveway with the tractor each day, to pull him onto the farm to pick up the milk from our bulk-tank milk cooler. Then he pulled Joe through the snow back to the main road. As bad as the winter was, there was only one day when Joe didn't make it to pick up milk, so we hauled the day's milk supply into our bathtub and put it in cream cans in the basement, storing it until Joe could pick it up.

It had been one hundred years since young Joseph Arens, his parents and siblings had set foot in Nebraska. Over that century, my own branch of the family had moved only sixteen miles up Bow Creek from where the Arens clan had started. We were still farming, like my great-great-grandparents had back in the nineteenth century, still struggling with weather and poor commodity prices. And our deep Catholic faith and the warmth of a friendly parish community were still the only things we counted on to make our lives meaningful and complete.

Haystack with a View

emember Curtis, you have to go out and get the eggs," Mom
would remind me as I walked through the door after school,
throwing my books down on the kitchen table.

"After the Flintstones, Mom," I'd reply. I liked watching the
Flintstones on our black-and-white television as a way to wind down
after a hard day in the first grade. In the early winter months, by the
time the Flintstones were finished in Bedrock, it was usually dark outside. As I donned my old cloth coat and overboots and grabbed the
yellow-wire egg basket, the only light in the farmyard was glowing
from the barn where Dad was milking cows.

I ran into the barn, patted the heads of several baby calves that liked
to suck on my fingers as I walked by their pen, and then walked to the
hen house for my evening chores. Several hens would peck at my
hands as I lifted them from their warm nests and gently lowered their
creamy-white eggs into the basket. I kept clear of that old rooster that
liked to fly up and beat me with his wings. A few pops from my BB gun
the summer before had earned me a little more of his respect.

After taking the eggs up to the house and soaking them in soapy
water, I called our fat old German Shepherd, Princess, and climbed to
the top of the tall haystacks along our manger where I knew Dad would
eventually come to pitch hay down to the cows. The dog was so heavy
that I usually had to push her up the high sides of the huge alfalfa

haystacks. There we'd lie, our backs sunk in the hay, Princess and me.

I watched my breath sift from my mouth into the cold, dark sky. I could see my whole world from the top of that haystack. Everything I knew was close, within shouting distance, and I was comforted by that knowledge. Princess and I thought a lot about God up there in our cozy nest. I could hear Dad talking to the cows as he finished milking in the barn nearby and wondered what Dad thought about God. I wondered what made those cows eat, give birth, and provide milk. I even wondered why God made that mean rooster that chased me.

Those cold evenings in the haystack with the dog are some of my earliest recollections of my life as a farm kid. I don't personally recall many of the antics my parents tell me about from my childhood. I have to take their word for it that at age two I would often stroll from our trailer house across the lawn to Grandma Caroline's house, knocking at her door, shouting what maybe was my first word, "cookie." I'm told that Grandma spoiled me. I only wish I remembered her more, because I feel a real admiration for her on many levels, particularly the example she set for her family about prayer.

I looked up above her bed and noticed for the first time a wooden crucifix that had probably always hung there.

Five years after Grandpa and Grandma moved into town, Grandma was diagnosed with cancer. One of the few memories I can recall about my grandmother is the last time I saw her. She was home from the hospital, and we went to visit her. When I came into the dining room, I looked through the doorway and saw her lying in her bedroom. Always a plump, kind lady, I thought she looked gaunt and tired.

"Come on in, Curtis. I won't bite," she beckoned me. I shuffled into the bedroom and sat beside her as she put her warm arms around me.

I looked up above her bed and noticed for the first time a wooden crucifix that had probably always hung there. There was a rosary beside her bed.

Mom's younger sister, Ceil, stayed with Paul and me so Mom and Dad could be with my grandmother more during her last days. I was riding my bike around the farm when Aunt Ceil yelled for me to come to the house. She sat me down on a wooden chair in the dining room, putting a hand on my shoulder. "Grandma Arens is in heaven now," she said. I didn't cry because as a six-year-old I didn't really understand what being "dead" meant. At Grandma's wake service and rosary the night before her funeral, I viewed her lifeless body, lying peacefully in a casket. I glanced down at her hands where her rosary was intertwined in her fingers, like it might have been when she was alive. I think it was the first time I'd ever been to a wake service or seen the lifeless body of a loved one who had died. It was an odd feeling, but it was not as frightening as I had imagined.

After her funeral Mass, by her grave at the cemetery, everyone prayed with the priest for the repose of her soul. I remember those around me reciting, "May the souls of the faithful departed rest in peace. Amen". We walked to the car in silence, and my front tooth—that baby tooth that had been wiggling and loose for a couple of weeks—fell into my hand. That tooth had worried me for some time. In my young mind, I couldn't help but thank Grandma for helping me lose it at her funeral.

Our farm was an inspiring place to grow up. Like every farm kid, I loved the seasons and still do. Springtime was calving time. My parents would pile my brother and me into our pickup nightly to drive to the pasture to check cows. Dad stood Paul and me up on the top of the pickup cab. We counted the cows and their baby calves to be sure they were all in the pasture.

But I think there was a more important reason behind those pasture rides. It gave us a chance to talk as a family. Those evenings in the pickup were like little family meetings. We often parked the pickup on a hill that gave us a sweeping view of our farm and all the land we were stewards over. We sat and talked about the land. Dad recalled stories of his own youth, and he and my mother discussed transition plans for the farm. Paul and I were only youngsters, but our parents were already thinking about passing on the legacy.

It was never assumed I would farm, but from a very early age I realized that no matter what else I did in life I wanted to live on our farmstead. The lure of the land of my forebears was too strong for me to ignore.

From our earliest days of childhood, our parents taught Paul and me how to pray. Dad tucked us into bed at night, kneeling by our bed to say the Guardian Angel prayer, a Hail Mary, and an Our Father before our heads hit the pillows. As we grew older, we added an Act of Contrition to our nightly prayer list. Paul and I learned about prayer by listening to my parents' stories of my grandparents. As Mom and Dad's own faith was being transformed and deepened through their involvement in Cursillo and a prayer group with several close neighbors, they taught my brother and me not to just pray for rain or for protection for our crops and livestock from the weather. Mom and Dad wanted us to pray that no matter what happened we would survive happily as a family, not only financially but also spiritually. Praying was changed from a direct request line to God for things that we wanted to a more mature type of conversation that recognized what we wanted for ourselves might not always be God's will for us.

When I first understood this concept, my prayer life and maybe my faith itself became something much deeper. I still find it extremely difficult to just turn things over to God. I want more control. But I have internalized the idea that God has a plan for each of us. I've come to appreciate every day I have on this earth a little more because of it. That's probably why my grandmother, who lived through drought and pestilence during the 1930s, felt so strongly about prayer, thanking God especially for the gift of food on the table.

I have learned from personal experience that no harvest is guaranteed.

Our daily bread on the table, like manna from heaven for the Israelites in the desert, was truly a miracle in her eyes.

Growing up, I heard my father repeat Grandma's words often: "A family that prays together, stays together." Those weren't just words to her; they were the hallmark of her life. I doubt if my grandparents ever started a meal without saying a table blessing. It is only fitting. Farmers, above all others, recognize no meal should be taken for granted. They know about God's miracles of a tiny corn or oat seed, like the mustard seed of the Bible, landing on fertile soil and bringing forth a bounty.

I have learned from personal experience that no harvest is guaranteed. I recall one extremely humid July when I was about ten years old. A beautiful, tall, ripened field of oats south of our place was ready to be harvested. The evening before we planned to combine the field, however, a gigantic, black-green wall of clouds rumbled in from the southwest. The temperature dropped 30 degrees in a matter of minutes. We sat in the house, watching the storm and praying the rosary, because my parents knew they were clouds carrying hail. We were jolted from our seats when a hailstone suddenly struck the kitchen window. A few stones hit intermittently, like someone was throwing rocks at the windows. Then suddenly, a foray of hailstones pummeled the ground and the crops, as gale-force winds raised the storm to a crescendo. Hail and wind smashed through windows, pounded dents into vehicles and tractors, ruined shingles and siding, and bruised livestock. I can still see my Dad walking out on the front porch to survey the damage. He looked out on what might have been a bumper crop of oats only ten minutes before. Hail covered the ground like drifts of snow. Tree limbs and leaves littered the farmyard. The oat crop laid flat in the field—a total loss. One look at Dad's sad expression told me the entire story.

Those were turbulent years on the farm in terms of weather. For a farmer, rain means life, because crops in the fields and pasture for live-stock depend on it. Half of my childhood was spent praying for rain, and the other half was spent praying we wouldn't be hailed out. During drought years, when mid-summer temperatures soared above 100 degrees every day and corn wilted in the fields, if a promising cloud would build or a few sprinkles would drop we often would sit as a family on the front-porch swing and pray.

One particularly steamy summer, we hadn't had any measurable rain for a couple of months. Aunt Ceil was visiting for a few weeks. A prom-ising cloud appeared in the west, and we ran out on the porch to watch it develop. A cloudburst dumped on us as it rained for several minutes, dropping an inch of desperately needed precipitation. After the storm, the sun broke through and an enormous rainbow bent across the sky from Bow Creek, over the farm, all the way to the northeast horizon. We hadn't seen mud for so long that my fun-loving aunt dragged me outside and we both sat in the mud puddles and played in the water left behind. No World Series winner or Super Bowl champion could have been happier than we were at that moment.

Another evening, Paul and I were feeding chickens when we noticed a dark cloud of dust forming in the southwest, over where Dad and our hired man, Robert Schaefer, were working in a field about a mile away. Mom hurried Paul and me to the southwest corner of the basement, while she searched for her parents, my Grandpa and Grandma Bickett, who were living at the time in a trailer house on our farm. They always took a walk to the pasture north of the farm about that time of night. As Paul and I huddled in the corner of the basement, we could hear the wind roaring outside. I rushed up the stairs and grabbed the family Bible, which I always considered a treasure, and ran back down to sit beside Paul. We both were thinking about Dad and Robert exposed in the open field. We started praying the rosary.

Mom found Grandpa and Grandma dodging falling branches near the woods north of the farm and hurried them to safety. Then she came to check on us. When the storm finally blew over, Dad and Robert safely returned home with a harrowing story. Dad had ducked beneath the tractor, hanging on to the wheels as the storm blew over. Robert took shelter inside an old, creaky corncrib. When Dad returned to the end of the field, he found large sheets of metal siding from the neighbor's barn surrounding the pickup. Amazingly, not one sheet had struck the truck.

Fluffy snowflakes began to fall, softly coating everything. I was grown now, feeding cows one evening in our hay lot, where we pile all of the big round bales of alfalfa and prairie hay for the winter so they are accessible to feed the cows through the snowy months. Our farm dog jumped up effortlessly onto a long row of bales. I paused from my work, stepped off the tractor, and leaned on one of the bales near where the dog was playing. She sat down beside my arm, and I stroked her ears. I recalled those early evenings when I was a kid, lying on my back in the hay and pondering life and God from the majestic top of a haystack. Looking into the dog's dark eyes, I thought about what my parents and grandparents have taught me about God and family.

As I stood in the snow, I glanced over to the house, lit softly against the misty dusk. I prayed to God that night that I might be a good and faithful husband and father, as I imagined St. Joseph was to Mary and Jesus.

Sometimes, I think I've failed in those departments. I'm too impatient with my kids. I raise my voice with them. I don't hand on my faith as clearly and as regularly as I should. When I'm tired from fieldwork or worried about hailstorms, drought, or what always seems to be impending poverty, I'm often short and terse with my wife. I procrastinate, especially with jobs around the house, and I know it drives an organizer like Donna absolutely crazy. I don't work hard enough. I'm not as jolly as my grandfather was or as pious as my grandmother. I often feel inadequate in my role. But I ask the Holy Spirit to guide me. I resolve in my heart to just keep trying to do better.

I confide all of these insecurities to the dog, because she's such a good listener. The young pup wags her tail and pants in my face, eager enough to forgive me my trespasses.

Catholic School Confessions

Before you get on the bus tonight after school, stop by the convent," my third-grade teacher, Sister Veronica, told me that morning. "I have something for you."

I was curious about her request. I couldn't imagine what she could possibly have for me. As an elementary student at St. Rose School, I was quite shy and ordinary. I liked to read and was just learning how to write in story form. After one special excursion to a threshing bee in nearby Niobrara, where the sights and sounds of old-time farming methods—such as threshing grain with gigantic old steam tractors—had excited me, I wrote out in longhand a tale of the adventure called simply, "The Olden Days." It wasn't for an assignment but just for fun. I realized, even as a third-grader, that writing down my observations could be rewarding.

Nearly all of the teachers at St. Rose School at that time were nuns. School Sisters of St. Francis had taught at the school in Crofton since it opened in 1911, just as they had at so many rural Catholic schools in northeast Nebraska. Only a few lay teachers had been added to the staff. The strict and stern nuns of Catholic schools are notorious. At St. Rose, though, I found our Sisters to be friendly and compassionate.

Sister Veronica was our teacher for the second and third grades and the instructor who initially taught us what it really meant to receive Christ in the Eucharist. As we practiced to receive our First Communion, she taught us to fold our hands perfectly as we walked up to receive the host. But Sister Veronica taught me more than Eucharistic etiquette. Her instruction became for me the basis of a lifelong devotion to the Eucharist and an appreciation for the real presence of Jesus in the bread and wine.

"When Father holds up the host during consecration," Sister told us, "I always bow my head and repeat in my heart, 'My Lord and my God,' the same words Thomas said when he was asked to place his fingers in Jesus' wounds after the Resurrection." I've been repeating those words in my heart during consecration at Mass ever since.

"I know this book is a little advanced for you right now, but I think you'll like it anyway."

At the end of that school day, I rushed to the convent and knocked at her door. She appeared, holding a book in her hand. "I know this book is a little advanced for you right now, but I think you'll like it anyway," she told me.

It was an illustrated biography of Abraham Lincoln. I guess she saw some foreshadowing of my lifelong interest in history, even when I was in third grade, or perhaps my passion for history was born because Sister Veronica took the time to cultivate that in me.

She also is the teacher who offered many St. Rose students at the time, including me, a first opportunity to lector at daily Mass. Looking over the lector list at my parish today, many of them probably first learned the ministry while attending St. Rose.

When the School Sisters of St. Francis hosted a centennial celebration of their order at our parish several years ago, a number of nuns who formerly taught at St. Rose School returned to Crofton. Sister Veronica, well into her nineties at the time, was still serving a parish in Petersburg, Nebraska, as librarian. I was elated to see her, introduce her to my wife, and thank her for what she had done for me.

The School Sisters were not the only influences at St. Rose School on spiritual and character development. Father Dan Galas—a short, jolly, cigar-smoking Polish priest—became the parish pastor when I was in the second grade. We looked forward to the days when Father Galas came into the classroom to teach religion. Not only was he funny, often wearing a shirt with smiley faces all over it, but he taught us about the history of the church in a way our young minds could absorb.

At recess time, Father was the pitcher for both teams during our softball games. When someone would get out of line, heckle a poor batter, or argue over a close call at home plate, Father stepped in with a smile and delivered judgment with fairness but finality. He preached kindness and fair play by practicing these very virtues on the playground, in terms his students could understand. He blended catechism with games and faith with fun splendidly.

Growing up in a German family, Dad had taught us how to make the Sign of the Cross, reciting the words in German. The abbreviated German slang amounted to "Vater, Sohn, Heilig Geist, Amen," which loosely translated into, "Father, Son, Holy Spirit, Amen."

Father Galas, so proud of his Polish roots long before Polish Cardinal Karol Wojtyla became the beloved Pope John Paul II, taught us how to make the Sign of the Cross in Polish. We loved listening to him recite common Catholic prayers, like the Lord's Prayer, in his native tongue. Although this was after Second Vatican Council and altar boys and Catholic students were no longer required to learn prayers in Latin, Father Galas delighted us by reciting the Sign of the Cross and familiar Mass prayers in Latin as well.

He taught us there are a multitude of ways to express faith and that the culture, traditions and rituals of our Catholic Church are rich, worldwide and universal. By the end of third grade, several of the boys in my class had resolved to be priests. We wanted to be like Father Galas, who seemed so happy in his vocation. I suppose that zeal wore off as we all grew older and became more interested in girls than in a celibate vocation in the priesthood. But when I think of what I consider the depiction of a happy, well-rounded, faith-filled rural parish priest, I often think of Father Dan Galas.

One of the perks for a middle school student at St. Rose was being chosen to serve as an altar boy for funeral Masses. Usually, if you were related to the deceased, you were the first chosen. Just as often, the family of an elderly parishioner with no family members of altar boy age didn't mind if a random sampling of the sixth-, seventh- or eighth-grade boys served for the Mass. Not only were we excused from class during the Mass, but we also rode with the pallbearers and the priest in the mortician's car to the cemetery service.

Dick Wintz, our local funeral director, always went out of his way to visit with us as we rode to the cemetery. We enjoyed the extra attention, and deep down we were honored by the sense of responsibility in being a part of one of the last acts we could do for someone who had passed on, particularly the veterans who were memorialized with an impressive military graveside ceremony.

Although funerals were often sad events for families of the deceased, there were lighter moments. Much of the difficulty for altar boys during funeral Masses came from trying to get incense to light properly. We always breathed a sigh of relief once it was lit and Father had blessed the casket with the sweet-smelling smoke. Normally, one of the Mass servers had to take the incense out of the church immediately after Father had blessed the casket and extinguish the lit embers. During one particular funeral Mass when I was serving with another classmate, we forgot to put out the incense and instead placed the burner back in a metal holding cabinet near the sacristy, just off the main sanctuary of the church.

During Communion, a cloud of sweet, white smoke billowed out of the room where the cabinet was located, filling the church. One of us rushed into the room and took the incense outside, because we knew if Father noticed the smoke we would be in for a good lecture about properly carrying out the responsibilities of serving Mass.

I also recall serving a Saturday evening Mass as a seventh-grader, complaining with other servers before Mass started about Nebraska's Cornhuskers losing another close college football game to archrival Oklahoma. Nearly all Nebraskans are "crazy" about the Huskers. Growing up, I enjoyed watching them play on TV whenever I could, but more often I listened to the game on the radio while working in the field.

On this particular night, a visiting pastor was preparing for Mass when he overheard our discussion. He quickly replied, setting everything perfectly into perspective, "Boys, you just have to remember that Oklahoma isn't number one. God is."

When I was in the sixth grade at St. Rose, my mother came home one semester from parent-teacher conferences with a scowl on her face. She showed me the report card she'd received. Mr. Filips, had given me a "C" in science. It wasn't the "C" letter grade that frustrated her. He had told her that I wasn't working up to my potential. To my mother's credit, she didn't blame Mr. Filips. She laid that responsibility firmly on my shoulders, where it belonged. I buckled down in that class and improved. Mom didn't care what the grade was, but she did care that my brother and I always did our best. As I look back, I learned more from that "C" than I had from all the other "A's" and "B's" I'd earned along the way.

I give my Catholic education at St. Rose School the credit for the man I've become. I appreciate Sister Veronica for showing me how much fun American history can be and thank many of the other School Sisters for having faith in a very shy and awkward farm kid. I acknowledge Father Galas and other priests who served our parish so well during my formative years for my positive, open mindset about my faith and for at least posing the question of a religious vocation at an early age.

St. Rose School offers education for grades one through eight, so students typically attend kindergarten and high school in the local public school. My oldest daughter, Lauren, for example, attended kindergarten at the public elementary school across the street from St. Rose.

We have a good public school in Crofton, with caring educators. But I admit freely that it was one of my happiest days as a Catholic parent to hold my daughter's hand as we crossed the street from her kindergarten classroom one winter afternoon so she could attend the first grade roundup at St. Rose School. Registering her for classes for the following school year at St. Rose, where I first was affirmed in my faith, was a wonderful feeling. To have the opportunity for a private, Catholic education for my children in my own hometown is a fantastic privilege.

Knowing my children are being taught the richness of our faith and expected to learn responsibility and values is so sacred to me because I know how difficult it was for the founders of our parish to build the school in 1911. At the time, I'm certain many people wondered why the new public school in Crofton wasn't good enough for those early Catholic families and their children. But in many public schools then, the King James Bible, widely accepted by most Protestant denominations, was sometimes taught in the classroom. Public schools of the day, in some places around the U.S., utilized decidedly anti-Catholic references in literature and history courses, using books that referred to "deceitful Catholics" or "vile popery." Many Catholics at the time became staunch supporters of separation of church and state because they wanted the freedom to build their own Catholic schools and teach their children their own Catholic identity without interference.

The German families settling in this area had been persecuted in Germany because of their Catholic faith, so it was worth every sacri-fice to them to be able to offer Catholic edu-cation to their children. I'm also reminded by the old-timers that if there hadn't been Catholic schools in some rural areas many of the farm kids would never have gone to school at all. The School Sisters may have been stern during those early years, but they were also dedicated, learned women who offered quality education in reading, writing and math, as well as religion. Knowing this heritage and the sacrifices made on my behalf makes it easier for me to lay it all on the line—"time, talent and treasure," as the current St. Rose pastor, Father Joseph, puts it—to keep our school alive for my own children and the children of my friends and neighbors.

> *"The reason I sent my kid to St. Rose was because Jesus is in the classroom there every day."*

One day a friend and I were leaning over the box of my pickup, dis-cussing the importance of the Catholic school to our community. "The education is important, don't get me wrong," he said. "But the reason I sent my kid to St. Rose was because Jesus is in the classroom there every day."

Catholic schools are not perfect. No single institution, even one operated in the name of Jesus, is perfect. It takes valiant effort on the part of teachers, pastors, administrators, parents and students to con-tinually strive for excellence. I'm guessing that some of the students who struggle with coursework or detest religion class the most will be the ones who will look at their St. Rose education thirty years down the road and give thanks they attended a school where prayer was not only allowed but encouraged.

I realize that a Catholic education is not a guarantee our children will remain active in their faith as adults. Future spouses, in-laws and friendships greatly influence how a person keeps the faith. Enrolling my child in a Catholic school doesn't rid me of the responsibility to impart faith in the home. But I think a Catholic education, combined with a faith-affirming spirit by parents, produces students who have a strong, central faith and character base to start from—and that is half the battle in this world, in my opinion.

After Lauren and other future St. Rose first-graders met their teacher, Mrs. Blaalid, the principal, Sister Charlotte, and Father Joseph at the first grade roundup and finished their classroom activities, everyone stood around the hallway, enjoying treats provided by the school board members. Lauren skipped up to me and showed off some artwork she'd created.

"Well, Lauren, how do you feel about going to school here next fall?" I asked, wishing Sister Veronica could meet my daughter.

She just flashed a bashful smile. We headed out the door and walked toward our van. "I can't wait," she whispered.

I squeezed her little hand gently. "Me neither."

Be Not Afraid

I t is difficult for a teenager to understand the fragility of life. In my teenage years, I watched some of my closest elderly relatives wither from age and fatal illness, yet I felt as if I could do anything and thought I would live forever. Death was the last thing on my mind. Still, as I witnessed my grandparents' and aunts' and uncles' anguish and watched them handle their suffering in a most accepting, faith-filled way, I learned an important lesson that I have only come to appreciate later in life.

Just months before my Uncle Ab was diagnosed with a cancerous tumor at the base of his spine, he and my dad's sister, Aunt Ceil (my mother also has a sister named Ceil), asked me to join them on a fishing trip to northern Minnesota. Living and farming just across the field from our family, Ab and Ceil were the ones who babysat me as a youngster. Their children were more like brothers and sisters than cousins to my brother, Paul, and me.

I often fed Ab and Ceil's chickens and did chores for them while they vacationed, and they wanted to return the favor by taking me on a trip. It was the first real fishing trip I'd been on, if you don't count fishing for bluegills in a local farm pond or for carpsuckers in Bow Creek. I looked forward to a week of fishing and, if the truth be known, getting out of the hot, dirty job of piling straw bales into the barn back home. But what really turned out to be the importance of that week of fishing for walleye, northern pike, and perch in Minnesota was that it allowed me to experience a special time with my aunt and uncle. That is a blessing I have carried with me my entire life, because of what happened next.

That fall, Ab's back began to bother him. I remember that at one point Paul and I helped Ab deliver newspapers on a Saturday afternoon on a route north of Yankton because his back hurt him so much he couldn't reach out of his car window to place the newspapers in the boxes.

Ab was diagnosed with cancer, and within a few months he was bedridden. There was very little doctors could do to treat his disease back then, so Ceil cared for him at home as best she could. She tried to keep him relatively comfortable, but listening to his terrible cough all day and all night had to have taken its toll on her.

At first Ab was both angry and sad, but he gradually accepted his fate.

Dad was then a Eucharistic Minister at St. Rose Church. On many Saturday evenings after Mass, we brought Communion to Ab and Ceil and then stayed to visit. During those evenings, I heard many stories about farming along Bow Creek that have stayed with me. I also heard stories of family and faith that have molded my own view of these things.

At first, like anyone who has received a death sentence, Ab was both angry and sad, but he gradually accepted his fate. Both he and Ceil knew he wouldn't survive this illness. He withered away over time, and he was to endure a painful sort of death. But one Saturday evening when we visited, Ab was completely pain free. That night, we stayed longer, laughed and visited as we'd done before he was sick. The next day, however, the pain was back with vengeance. By April, Ab was in the hospital and the family was called in to bid him farewell.

I was allowed to go into the hospital room shortly before he died. Ab was resting. His eyes were shut and his face looked pale and yellow. He didn't respond to anyone. We went home, knowing he would not make it through the night. Ab and Ceil's grandchildren—Bruce, Glen and Karla—and their parents Darrell and Pat—were staying with us. We had all just gone to bed when the bedroom door creaked and Dad peered into the room and said quietly, "Boys, Ab is with God now."

A short time later, the extended family all came to our house and sat around our kitchen table sharing tears and memories of Ab. The adults drank a little and laughed a lot through that long night, and they began to plan his funeral. I thought at the time that this was a wonderful way to honor Ab: by celebrating his life on earth and his entering eternal life at the same time.

After Ab's death, Ceil went on with her life, but it was obvious to me that she terribly missed her lifelong companion. She had never driven a car on her own, relying on her husband all of those years, so she bravely started driver's education classes in town with several teenagers my age. There is something about that detail that still brings a tear—of joy, of sorrow—to my eye.

Two months after Ab died, I was working their fields to help out. As I finished, Aunt Ceil came from the house for a visit as she often did, but this time she didn't say a word. She reached up to the fender of the tractor and handed me Ab's old fishing tackle box, the one we had used in Minnesota. She smiled as I took the box and grasped it into my arms, utterly speechless myself, and she returned to the house. That tackle box is even today one of my most sacred possessions.

In early August, Ceil was practicing her driving early one morning with a driver's education instructor and wasn't feeling well. It turned out she was suffering a heart attack. After consulting with the doctor in Crofton, the instructor drove her as rapidly as possible to the hospital in Yankton, but it was too late. Ceil had slipped into a coma and died a few days later. In less than a year, a couple I had loved and cared for deeply was gone from the earth. Thus did I first experience the universality of the human reality of mortality.

Grandpa and Grandma Bickett were still living on the farm near Howard, S.D., when I was a kid, but they visited us quite often. Grandpa was a quiet, gentle farmer who could play instruments by ear. He entertained us by sharing his Santa Claus voice and telling stories about his World War I experiences.

Grandma was a great storyteller, a blessed gardener, and a wonderful cook. During one visit to Crofton, Grandma became quite ill. She was diabetic and had already suffered a stroke a few years earlier. When she visited the doctors at Sacred Heart Hospital in Yankton, they decided to do exploratory surgery. Midway through the surgery, one of the doctors left the operating room to talk with family members, shaking his head. Cancer had infected Grandma's intestines, and he wasn't sure there was anything he could do. He told everyone he would try his best, and he went back in and surgically removed much of Grandma's small intestine.

At that time, Grandpa and Grandma Bickett decided to make a life change. Grandma Catherine needed to be closer to her doctors in Yankton, so the two of them moved into a trailer house on our farm, near the spot where my parents and I had lived when Grandpa and Grandma Arens were living on the place. After several months, Grandma Catherine recovered from her cancer, and her prognosis for the future looked bright. For Paul and me, having our maternal grandparents so close to us during our formative years was a true blessing. We attended weekend Mass with Grandpa and Grandma. We enjoyed going to ballgames in town together. When I hunted my first rabbit—a rite of passage for a farm kid—Grandpa helped me clean the animal and Grandma cooked it with mushroom sauce.

On weekends when Mom and Dad had prayer group or Cursillo activities or were attending a neighbor's wedding or anniversary party,

Paul and I often stayed overnight with Grandpa and Grandma. We knew we would awake to a breakfast of thin buttermilk pancakes and thick sausage patties. Paul and I learned about courage by getting to know our grandparents so intimately. We watched Grandma inject herself with an insulin shot every morning and then go out all day long as if she had just received a new lease on life—gardening, visiting, and living life to the fullest.

We enjoyed gardening with her and Grandpa, because they often engaged in a lively banter back and forth.

When we seined minnows from Bow Creek for a neighbor's bait shop, we often saved the two-pound carpsuckers that got caught in our nets for Grandma to clean and fry. We enjoyed gardening with her and Grandpa, because they often engaged in a lively banter back and forth. Grandpa would "accidentally" spray Grandma with the water hose while watering flowers, and she would playfully scold him. Because my grandparents were living on the farm with us, I also became very close to our cousins who visited the farm and stayed for days and even weeks during the summer months.

Throughout those years, when Grandma was living a full life again after beating cancer, she became devoted to Our Lady of Lourdes. She purchased holy water from Lourdes through the mail and daily placed a drop of Lourdes water on her tongue.

Grandma also nurtured a love of reading in her grandchildren. Paul and I were commissioned each Saturday to go to Eastern Township Library in Crofton and pick out some new book, preferably a pioneer story of some kind, for Grandma to read. She devoured books. She particularly loved the "Little House on the Prairie" books by fellow South Dakota native Laura Ingalls Wilder. Grandma was personally familiar with the towns and lakes Laura described so vividly.

Grandma's cancer returned ten years later. When I was a college freshman, she became quite ill. My grandparents were forced to move into the main house with our family because Grandma could no longer take care of herself and Grandpa.

During that time, my mother began a new vocation in life, which was being the primary caregiver for both her two teenage boys and her own parents. I guess they call it the "sandwich generation" now, but back then it was just the way things were. While my brother and I were growing up, Mom's job was taking care of the farm records, helping on the farm and in the garden, and cooking and cleaning for our family and a hired man. She did all of these jobs well. Carrying on a tradition from her mother, Mom loved making homemade bread and cookies. She loved vegetables and tried, sometimes to her frustration, to get her sons to eat them as well. Also like her mother, Mom loved books and served as our mentor when it came to studies in school, especially research and book reports. She encouraged us to explore our talents, urging me to continue to write and Paul to learn to play the piano as well as another instrument in school band. When either of us boys had a crush on a girl or were depressed about something, Mom was always there with a good ear and sound advice.

When Grandma became ill, Mom stepped up to help in her mother's long struggle with the recurring cancer. If Grandma was sick in the middle of the night, my mother answered the call. I'm sure it was difficult for Mom to watch Grandma wither away, but I think she called upon all of her faith to make some kind of sense of Grandma's plight. She always told us we were all blessed because after that first surgery we had all been allowed time to be close to Grandma for several more good years.

After Grandma died in 1984, my mother continued to care for her father. Although Grandpa Arnold suffered through long, lonely days, sitting in his chair in our living room playing old tunes on his harmonica, he appreciated the presence and love of my mother and all of his children and grandchildren who visited him.

Grandpa's mind began to wander, and he sometimes became disoriented. One weekend when I was home from college, Mom and Dad were attending a wedding in town. I needed help caring for baby pigs in the hog barn. I asked Paul to come from the house, where he was sitting with Grandpa, to help me get a sow that was birthing outside in the pen into the barn. We were gone only a few minutes when we heard a loud banging on the front door of the house. We ran to the house and opened the door to find Grandpa, upset and angry, trying to get out the door by beating it with his cane. He thought we had locked him inside the house. He couldn't recall us telling him where Paul had gone. Later that evening, it all came back to him and he apologized to my brother, but it was obvious Grandpa's mind was failing.

During his last couple of weeks of life, to my mother's dismay, we had to move Grandpa into a nursing home in Bloomfield, Nebraska. He was obviously sad about the entire experience, although he didn't really understand what was happening to him. When we visited him there he often did not recall our names, but if we sat beside him to pray he could always say the rosary with us. I love spontaneous prayer, talking to God from the heart. Catholics are sometimes chastised for reciting memorized prayers, without feeling the emotion or connection that prayer should foster. But for Grandpa, memorized prayers like the Lord's Prayer, Hail Mary and Apostle's Creed were the only prayers his failing thought processes could muster. He may not have known much else, but he never forgot how to communicate with God.

After Grandpa died in 1986, Mom was lost in many ways. She had been caring for her parents so intimately for so long that their deaths challenged her basic sense of purpose around the farm. About this time, another of the periodical farm crises had ensued. Land values and grain and livestock prices had plummeted, so relationships with lending companies had become strained. Many farmers were forced to quit farming or to take off-the-farm jobs to supplement their income.

After one particularly stressful meeting with a lender, my parents were told by their loan officer that maybe their son in college (me), who was planning to come home to farm, should consider pursuing a new profession. They were devastated.

Both my parents were determined to keep farming and pass the farm on to me, so they joined an interdenominational farm crisis group that met in the Methodist church in nearby Bloomfield. These meetings helped area family farmers share their sadness and frustration and look together as a group to God for some hope. To help make ends meet, Mom started cleaning houses in town. Then she took a job at the nursing home in Bloomfield where Grandpa had lived.

In many ways, her work as a home health aide is a life-affirming prayer for Mom.

Mom has since blossomed in her third career (after being a lab technician and a homemaker), learning to care for the elderly and others who need a hand as a home health aide. This is a job where she can use the loving techniques she learned by caring for her parents in their last years and minister to the families of those who are sick and dying. Having lived through the same experiences as many of her clients' families, she can pray with them as someone who has been through it before. In many ways, her work as a home health aide is a life-affirming prayer for Mom.

When I was in high school, I was with my family at the annual summer bazaar at St. Joseph's Church, playing some outdoor games on the lawn outside the old school basement, when a friend asked—to my great surprise—if I was going out for football in the fall. Never having considered myself much of an athlete, I was taken aback by his question. I was already going to be a junior, and I had never gone out for a team in any sport (except the junior high basketball team at St. Rose School). But I should thank that friend today, because I did consider it after that night and talked it over with my parents. I worried about the time practice might take away from helping Dad with harvest, and frankly I was worried I wouldn't be any good. During Communion time at Saturday night Mass, the congregation was singing that familiar hymn based on Isaiah 43:1-3: "Be not afraid. Know that I am with you always. Come follow me, and I will give you rest."

As I was singing I couldn't help but think about how afraid Ab must have been when he learned he wouldn't make it. My grandmother, too, knew real fear of dying, but faith in Christ lifted her up through her most difficult times of suffering. In retrospect, trying out for football seemed like a miniscule matter to be pondering so intensely. They had to deal with death, and I was dealing with an activity that is fast-paced and full of life.

I decided to give football a try. To my surprise, I loved it. The sense of belonging to our team and fitting in—although we had a dismal, losing season—made me look forward to my senior year. I started off my senior year playing varsity center on offense and tackle on defense.

We don't have a Catholic high school in Crofton, so although the team was born from a public school setting many of the Catholic kids on my team made a ritual of going to St. Rose Church across the street from the high school before each game. We'd light candles and take time to pray. Many of the other Christian players on the team started joining in. And win or lose, after our games those who wanted to often gathered in the end zone near our busses, took a knee, and prayed the Lord's Prayer together. No one was coerced into this act, and it wasn't organized or even encouraged by any of the school coaches or staff. We kids did it because we had something to say to God, giving thanks not only for what turned out to be a very successful season but also for giving us the opportunity to play and sparing us all from serious injury.

During those teenage years, I was learning multiple lessons from those around me by participating in activities such as football, FFA, 4-H, and the school yearbook staff. But I learned the most about life by witnessing several valiant, faith-affirming deaths in my family. I find that curious, even today, but as I grow older I understand that death is the great leveler, the experience that makes us human beings come face to face with our mortality and our God.

Keeping the Faith

I threw the last bag into my chocolate-brown, 1972 AMC Ambassador sedan and hopped behind the wheel. I rolled down the window and waved to my parents. I was leaving the farm for Lincoln to start my studies at the University of Nebraska, fulfilling a long-time dream.

I'd been in Lincoln many times—on family trips, for state FFA conventions, and during 4-H activities—and had fallen in love with Lincoln and, particularly, the university campus there. I would be living in a small dorm two miles from the downtown main campus on East Campus. It was sometimes called "Dirt Campus" because it was home to the College of Agriculture and many of the students who actually lived there came from farms and small towns like Crofton, my hometown.

Like most young people, I didn't think much about what my "big step" toward college meant in the lives of my parents, particularly my dad. Our hired man, Robert Schaefer, was still working for my family. It was like having another boss on the job and in the field. He was a part of our family. Yet Dad was still going to miss me at work.

I had never driven a car by myself through any town larger than Yankton. I had never driven by myself for over three hours. I had never encountered any traffic to speak of, except maybe a herd of cows being moved on the road. I had never lived alone before. I hadn't even been away from home all that much, except for that fishing trip with Ab and Ceil.

I glanced over my shoulder as I drove off. Mom retreated into the house, but Dad stood on the sidewalk. He looked sad, watching my car drive away. I was scared and ecstatic at the same time. I hadn't even left the driveway and tears streamed down my face. What was I getting myself into? How in the world would I survive in Lincoln, not knowing a soul? I really wanted to turn the old car around and stay home on the farm, but I knew that I had to start this new chapter in my

The tears kept rolling until finally, about a half-hour later, I grew calm.

life right away, even if it was a step into the unknown. The tears kept rolling until finally, about a half-hour later, I grew calm. I kept driving into the unknown.

Fortunately for me, my first college roommate, Ron Kulwicki, was an upperclassman, a farmer from a small town, and a willing mentor for a naïve freshman like me. We hit it off almost immediately. Ron explained how to find the things I needed and how to get along in the classroom and on campus.

He held several part-time jobs, so I had a great example of industriousness. With only a couple of small scholarships to help pay my way through college, I eventually had to pick up three part-time jobs on campus as well. Ron also was Catholic, so I had someone to attend Mass with right off the bat. We usually attended church at St. John the Apostle, a large, family-oriented parish near campus in east Lincoln. The priests there welcomed Catholic students from UN-L with open arms, and we became acquainted with some of the parishioners. We also went to Mass at the campus Newman Center downtown, whose chapel is absolutely beautiful. But for many of the Catholics on Dirt Campus, St. John's was more convenient and more comfortable because it had a familiar family spirit.

I was fortunate to make so many Catholic friends early in my college career. College is often the time in a person's life when it's easy to fall away from the Church, when faith doesn't seem to be a priority. To be honest, because of a good support team around me, I didn't think of attending Mass as a burden. It was such an important part of my weekly ritual that if I had missed a weekend I would have felt like I had missed out on an activity I depended upon for survival. By my senior year, many of us still living in the dorm often asked incoming freshmen whom we knew were Catholic if they would like to carpool to Mass with us. I shared in mentoring the faith of these new students, just as Ron and many others had shared their faith with me when I first came to Lincoln.

In high school, whenever I had vented frustration over high school social activities, friendships, or romantic pursuits, my mother always insisted that college would be much different. "The people you meet in college will be the people you associate with for the rest of your life," she told me many times. I didn't believe her, but she was right. I remain in touch with many classmates and friends from high school who returned, as I did, to our hometown to raise their families or who visit for alumni and church gatherings. Yet the friends I made in college— with whom I shared a dorm, cafeteria meals, new romantic interests, and breakups as well—remain my closest confidants two decades later.

I think friendships, good or bad, can make or break a college experience, particularly for a quiet freshman like me. Early on, I was so reserved that I worried that I wouldn't make friends at all. One late Saturday night, I was sitting in my dorm room studying. Ron was working as a night campus security guard downtown, so I was alone. One of the upperclassman knocked on my door and asked if I wanted to come over to his room and "shoot the bull" with the rest of the guys. Sensing my insecurities, he quickly added, "Just stop by and have a few laughs." He was trying to help me fit in and get to know the other guys living on the floor. So I went over to his room and stayed up all night playing cards. That openness to me as a person by upperclassmen who had been on campus for a while was quite reassuring. To this day, I still appreciate that gesture, even though I can't tell you the guy's name.

The friendships established during my college years weren't confined to Saturday night bull sessions. I became involved in the university's 4-H Club. The club members were active in numerous community and college service projects, such as promoting an on-campus blood drive, helping Lancaster County 4-H'ers with speeches and demonstrations, and handing out candy to trick-or-treaters at the Lincoln Children's Zoo at Halloween. We were a lively bunch. We even had a singing group called the "Outreachers" with twenty members.

Now, I can't carry a tune in a bucket. One Sunday when I was in high school, I stood at Mass beside my Dad and we both did our usual best to sing along with the hymns. Afterward in the car on the way home, Dad turned to me, "You know, Curt, I always wished I could sing, but today I wished you could."

That didn't stop me from joining the Outreachers and having a splendid time singing for nursing homes, 4-H groups, county achievement days, and even for a national convention of Cooperative Extension educators in Omaha. We all enjoyed practicing at a church social hall near campus every Sunday night. We sang old favorites and familiar church songs like "Pass It On" and "You'll Know We Are Christians." Our song list had a decidedly Christian perspective, because nearly everyone involved was a practicing Christian, although we were members of several different denominations.

Ron graduated and got a teaching job. My new roommate, Keith Berns, also was a great influence and comrade during my junior and senior years. Keith was well grounded in his faith and over time we became quite close. Because of Keith, I started to come out of my shell.

He was able to drag me to campus and 4-H functions and dances. He and I donned homemade "Ghostbusters" costumes, modeling ourselves after the popular movie of the day, and were a huge hit with the kids at the Children's Zoo at Halloween. We took our show on the road, walking in costume down through stacks of books in the campus library and through neighboring dorms, asking other students—particularly the pretty co-eds—if they'd seen any ghosts. It was all in harmless fun and we drew lots of smiles around campus. But unfortunately I'm certain today we would be considered a campus security threat!

I hadn't realized I was so insecure and quiet in high school, but I felt for the first time a new self-confidence in college.

One of my cousins from back home moved into the dorm where I was living. One night at the cafeteria, after everyone else returned to their dorm rooms, she sat down beside me as I was finishing supper. "I hardly know you these days," she said. "Back home, you always seemed so serious. You hardly ever laughed or had any fun," she explained. "Now, you're out playing basketball with the guys, joking around here. It's really good to see you opening up." College life was growing on me. I hadn't realized I was so insecure and quiet in high school, but I felt for the first time a new self-confidence in college.

Keith initiated a dorm Bible study, which was held weekly in our room. We took turns planning the simple study program and the entire session usually lasted less than an hour. For those of us who were stressed out about academic and work schedules, exams, and the all-encompassing question of life after college, the Bible study was a welcome respite. We often crammed over twenty students into our tiny dorm room, reading Scripture, talking about life challenges, and laughing together. Our only rule was to never quibble over the differences between our faith denominations. It seemed that these were more important to our various church leaders than they were to us. In that room, at that time, we were all believers in Christ, and that was all that mattered.

I had grown up in a relatively large Catholic congregation, so I had little working knowledge of any particular Protestant faith, let alone Judaism or other non-Christian faiths. My parents believed that Christians were united in a faith in Christ, and there was never animosity between members of different churches in our community. In fact, in Crofton, the three churches always cooperate in ecumenical worship services around Thanksgiving, Christmas and Easter, rotating the service to a different church each time. That spirit of mutual respect and cooperation was the childhood experience I personally brought with me to college.

However, I learned at college there were strong elements of anti-Catholicism among some students. I attended what was advertised as a "Christian crusade" one evening at the city campus student union, where I heard a speaker demean Catholics in broad generalizations. I was stunned and walked out of the room, feeling entirely disenfranchised by the many formal Christian ministries on campus, other than the Newman Center. I found that misunderstandings about my faith were widespread. Even some of my close friends made lighthearted jabs at what they perceived as my lack of knowledge of specific Scripture passages. At first I was offended, but after visiting with some of my Catholic friends I realized that we knew all of the Scripture passages, the parables of Jesus, and the stories of the New and Old Testaments. We just hadn't learned them in the same way as our Protestant brethren: We knew the faith messages but not always the exact chapter and verse numbers.

This whole experience forced me to look inside myself and question for the first time why I was a Catholic. During college, if I had been dating someone for a while, I always asked if she would attend Mass with me, maybe as a litmus test for a future relationship. With some of my dates, faith became an important issue of debate. Girlfriends asked me why Catholics genuflected, why we believe in the real presence of Jesus in the Eucharist, and why we pray to the Blessed Mother. I had not thought about why I believed any of these things before. I just believed them without question. Now these questions floated in my mind, and I began seeking Scriptural answers that my friends could understand as a way to explain my beliefs.

One day while visiting a Christian bookstore, I noticed an absence of the traditional Catholic books, like booklets about the saints, First Communion gifts, and Catholic Bibles. I finally found a book that was supposedly about Roman Catholicism and began reading. The pages I read described the consecration of the host and other Catholic liturgical rituals as cult-like and demonic. Overwhelmed with anger, I stomped up to the clerk (who was probably just a student worker) and explained that the book I was holding did not describe the Catholic faith I knew and was objectionable material in my opinion.

The young woman smiled and told me she would tell the owner of my complaint, but of course she was aware—even if I wasn't—that the store was a "Christian" bookstore that did not include Catholics in the fold.

During my freshman year in college, Grandma Catherine withered away from the cancer that had returned to her body after a long absence. Whenever I returned home, I could see she had gone downhill. She was a shadow of her former self, with her skin clinging to her bones.

One night while I was studying in the dorm, Mom called to tell me that Grandma probably wouldn't make it through the night. A couple of hours passed and a second call came. Grandma was gone. I started driving home for the funeral, and memories flooded over me of the hours upon hours Paul and I had spent with her, eating her special pancakes, enjoying her cookies, and watching the Lawrence Welk or the Donny and Marie Osmond shows.

During the funeral Mass, I offered to deliver a brief eulogy. My hands trembled as I stepped up to the pulpit; my voice cracked as I began to speak. I couldn't summarize Grandma's life in only a few words, but I hadn't forgotten her prayer of suffering. I talked of her love of life and her devotion to our Lady of Lourdes. Finally, I tried to offer the comforting thought that as long as we remember the best of someone we've lost, as long as we try to emulate his or her life in our own, then that person never really leaves us. I stepped down from the sanctuary and stumbled back to the pew. I lowered my head. I felt I'd made a pathetic attempt at explaining how I felt, but afterward my aunts and uncles and parents thanked me for those simple words.

> *For me, the college experience brought me closer to my full expression of faith in Christ.*

Since then, I've been asked to eulogize my other deceased grandparents, as well as several aunts and uncles and even a friend who died tragically in a car accident. Each time, I feel like I'm an imposter, trying to soothe a loss that can't be reversed. Each time with God's help, however, I am able to get through it, and I hope that I have imparted some comforting thoughts each time.

Just two years later, Grandpa Arnold died, just before my college graduation. This time, the call came to my dorm room unexpectedly. Diane, another good friend who was almost like my "adopted sister" in the dorm, took the call and was the one who broke the news. I hugged her, and she listened as I recalled many fond moments I had enjoyed with Grandpa. When I returned home for the funeral, several of my college friends accompanied me for support. That's the way it was with our group.

For me, the college experience brought me closer to my full expression of faith in Christ. Because of the fellowship my college friends provided, I came to deeply love all true friends in the faith—those who may profess different views than mine in some respects but are willing to listen to and accept me as a Catholic Christian who is always in the process of forming my adult faith.

That support and challenge is something I wish for my children when they are in college.

Dad and St. Isidore

I pounded the rough, worn-out baseball in my glove over and over. I looked to the west, down our long quarter-mile driveway toward the highway, and glanced back into the old glove, with its leather laces falling apart. The sun was sinking slowly in the horizon when I noticed a cloud of dust at the end of the drive and the familiar hum of the old Farmall 560 diesel tractor. Dad was heading home from the fields. He'd been there since daybreak, planting corn with our four-row planter, up and down hundreds and hundreds of rows all day. Many times at planting, the cold, northwest wind almost blew him from the open-air tractor seat. Other times, scorching sun might beat down on him all day. I never realized, until years later when I was planting corn myself, how tiring it could be in the field all day without a break. On this particular evening, like many evenings about this time of year, however, I had another job for Dad when his fieldwork was finished.

I was in sixth grade and infatuated with baseball, but I wasn't very good. If someone threw a softball at me, about half of the time I'd either drop it or it would hit me in the head. I didn't have any older brothers to throw the ball around with me and I wasn't in little league, but I was tired of getting my cranium bumped and bruised. On the school playground, I was often one of the last kids picked when we were choosing sides for ballgames. I knew that I could play; I just needed more practice.

I waited for Dad to get home at night to play catch with me. He would climb down from the tractor, his face covered with dust and grease, except for a clean area where he had licked his lips. He wiped the dirt from his bloodshot eyes. "Let me get a drink of water," he'd say.

> *We often practiced long after dark, and I slowly improved my game.*

After taking a long, cold drink from a hydrant and washing his face and hands, he'd put on a ball glove, grab the ball, and say, "Let's go."

He threw the ball to me as hard as he could. He threw pop fly balls high in the air for me to catch. We took turns pitching and catching. He batted balls to me, and pitched while I practiced batting. We often practiced long after dark, and I slowly improved my game. I never thought about it at the time, but Dad never played ball as a kid and had never played organized baseball. He and Mom took Paul and me to the ball games in town on Sunday evenings as often as they could find the time, and he enjoyed listening to the Minnesota Twins play ball on WNAX radio out of Yankton, S.D., but he wasn't a ball player, just a dead-tired farmer who wanted to give his son the skills and confidence he needed to succeed in life.

Dad has always been known for his kind heart…and his ornery streak. He is notorious for playing April Fool's Day pranks, like this past year when he called Mom on her cell phone while she was at work, pretending to be a telemarketer. She hung up on him.

When I was younger, he would call me out of bed early in the morning to report the cow herd was running down the road. By the time I pulled my jeans on and hobbled downstairs to help, he was standing in the kitchen, laughing.

Like his own father, Dad can be quite jolly. He enjoyed several years when he donned a Santa Claus costume and drove around the country-side on cold evenings over the Christmas season, bringing bags of candy to neighbor children, shut-ins and friends. On several nights, my mother, Paul and I served as elves, packaging hundreds of bags of candy and fruit in preparation for Santa's outings. I served as the reindeer Rudolph, chauffeuring Santa in the dark of night to area farms in our old pickup. We often cut the engine and silently coasted onto the place, parking out of view behind a barn or shed.

Santa surprised many parents by knocking on the door, bringing gifts to their families unannounced. Dad loved hearing a child's gift wish list and enjoyed the shock expressed by the parents even more. Paul and I reveled in the holiday experience. When Dad put on the suit, he *became* Santa Claus and was nearly unrecognizable, even to his closest friends. The only downside was that the suit was very warm, so even on nights when the temperature outside had sunk to zero degrees or below, Dad wanted us to drive with the air conditioning on or with the windows rolled down so he could cool off between stops.

I have always admired the beautiful, stained glass windows of our church at St. Rose. The colorful depictions of several saints come to life from those tall windows. I most admire the window depicting a Spanish man standing by his plow, his hands folded in prayer. Today, even our youngest daughter, Taylor, can look up during Mass at that same window and point out that the man depicted is a farmer, St. Isidore. I always identify with St. Isidore of Madrid because he was a poor farmer and field laborer who was not exalted in life but is celebrated now around the globe for his life of simple faith and service to others.

When my Uncle Clarence died, his family held a sale of his tools and other items from his workshop. I spotted a worn old booklet called the *Novena in Honor of St. Isidore* piled on top of a box of other knick-knacks. There must have been several important items in that box, because it brought more than I wanted to spend. Afterward I approached the man who had purchased the box, offering him five dollars for the St. Isidore publication.

The booklet—which lays out prayers and a simple, eight-day family novena service—was published in 1954 by the National Catholic Rural Life Conference, an organization dedicated to celebrating rural life in cooperation with God's grand design. The booklet includes novena prayers for farm families and meditations for rural citizens, parishes and pastors, including the beautiful "Litany of St. Isidore," which is reprinted in the appendix of this book.

The booklet has become one of my prized prayer books, not only because it was owned by my uncle but also because it has helped deepen my devotion to St. Isidore. I found a blue prayer card inserted in the novena booklet that I taped to a mirror in our kitchen so I see it first thing every morning. This special prayer to St. Isidore is attributed to G.T. Bergan, bishop of Des Moines, in 1947: "O God, who taught Adam the simple art of tilling the soil, and who through Jesus Christ, the true vine, revealed yourself the husbandman of our souls, deign, we pray through the merits of Blessed Isidore, to instill into our hearts a horror of sin and a love of prayer, so that working the soil in the sweat of our brow we may enjoy eternal happiness in heaven. Amen."

Isidore and his wife, Maria, were generous servants of God. They had only one son, who died tragically in his youth. Isidore could have been bitter, but instead he and Maria turned their lives over to God. Isidore's generosity was legendary. One would think that someone with so little to give others materially could not be very generous, but Isidore and Maria shared what little they had with others who were even more destitute than they were.

St. Isidore is the model of someone who has his priorities straight. Very few real biographical facts about him are known, but the stories of his life that survive provide inspiration to farmers and others living a rural life.

Isidore, born around 1080, worked all of his life as a farm laborer near Madrid for his employer, Juan de Vergas. He typically rose early to worship at church, remaining there to pray for long periods. Jealous co-workers chided Isidore to Master de Vergas, saying he didn't do his work because he was always at church or on a pilgrimage to some Christian shrine.

One day Master de Vergas decided to check on Isidore, to see if the accusations were true. Hiding behind bushes near the field where Isidore was working, his employer noted the laborer was indeed late to the fields. When Isidore arrived and started working, de Vergas was astounded when he saw a second team of white oxen, led by angels, plowing beside Isidore, completing the work in half of the usual time.

My father didn't have the luxury of angels leading white oxen, but on many occasions he had reason to believe the Lord was with him in his farmwork. Dad always related to his family the need to pray daily wherever we were, as well as the need to be aware of the presence of God not only while attending Mass on Sunday but also in daily work.

> *From colossal, life-saving events to the routine holiness of his day-to-day work, Dad has always invoked the name of the good Lord.*

From colossal, life-saving events, like his survival in the field from a twister when I was a young boy, to the routine holiness of his day-to-day work, Dad has always invoked the name of the good Lord.

For example, to start our ancient Farmall diesel tractors, like the 560 that now powers our front-end loader for feeding hay and pushing dirt and snow, the driver must hold a glow-plug button for a minute or two to heat the piston chambers, allowing the diesel engine to start from compression when the ignition button is pushed. When I was learning how to drive that tractor, I never could gauge how long to push in the glow-plug button so the engine would crank over on the first try. Then Dad told me that he always prayed a Hail Mary and an Our Father while holding in the button, before hitting the ignition. It turns out that is the exact amount of time the piston chamber needs to warm up! I still practice this daily prayer routine every time I drive the old 560.

One time when that same tractor ran out of diesel fuel, Dad needed to start it again to finish grinding corn for livestock feed. But it is difficult to start a diesel engine after it has run empty of fuel. He changed fuel filters, pumped the air out of the fuel line, and tried a number of things to start the old tractor. Finally, with milking time nearing and more work to do, out of frustration with the situation, Dad yelled out, "OK, Lord, you start this old tractor." He hit the ignition one more time and it fired off immediately.

Having lived and worked with a farmer all her married life, my mother would say, with first-hand knowledge on the subject, that St. Isidore was a typical farmer. For example, just like my dad, Isidore often brought guests home with him to share their sparse meals. His wife, Maria, knew this of course, so she always kept a pot of stew simmering on the fireplace in their humble farm home. One day Isidore brought a group of beggars home with him for a meal. As Maria started serving the group, she realized there was no more stew in the pot. Isidore asked her to look again. Miraculously, there was food enough for everyone.

On the farm, my mother always had extra food prepared for noon meals in particular, because she never knew whom Dad might invite. On a whim, he might ask a relative he saw in town, or a feed salesman, or a friend to join us, and Mom needed to be prepared.

While we were growing up, my mother was able to find anything we had lost, including the proverbial needle in a haystack. It must be a motherly trait, because if my children have lost some special doll or a tiny trinket, they know to ask Donna to find it, not me.

My mother often had help in her endeavors. If we lost something, her response was always, "Say a prayer to St. Anthony." Her devotion to St. Anthony of Padua, the patron saint of lost articles, probably helped solve a number of mysteries that could have been devastating. On one occasion, a hired man working for my parents was in a field on a tractor, cultivating weeds from the corn. His billfold must have worked out of his pocket and fallen to the ground. He was distraught, because he carried a large quantity of cash in his wallet, but he laughed when Mom told him to pray to St. Anthony. We all said a special St. Anthony prayer. When the hired man went to look for his wallet, he was shocked to find it on the ground along the field border. It was a small miracle, because the wallet could have been driven over by the tractor or covered with soil by the cultivator.

The clunk, clunk of ears of corn being gathered into the machine and deposited in a bin on top of the combine I was driving was always music to my ears. It meant that God was good to our family that season. When the harvest is bountiful, big ears of corn make that familiar clunking sound as they enter the combine, where the grain is shelled from the cob in the bowels of the machine.

One day, I looked up and peered through the dusty glass of the combine cab to see Dad driving our rusty, green Chevy truck into the field, parking it on the other end. As I came to the end of the field, I pulled a little lever. The auger that deposited corn from the combine popped out over the empty truck box. Dad stood beside the truck, "eating" the grain dust coming out of the combine as I unloaded the bounty into the truck. I looked again and saw him stumble. He clung to the mirror attached to the side of the truck cab with his hand over his face.

"You are not fine. We have to get you some help."

He was coughing. He coughed and choked and staggered in the dust and dirt. I stopped unloading and turned the engine off. I leaped from the combine cab, jumped down the ladder of the tall machine, and came to his side. Dad continued to cough and struggle. I wanted to cough for him. I wanted to breathe for him. I didn't know what to do.

"Get the water jug," he whispered between coughing fits. His eyes watered. He fell, slipping to the ground. He gulped from the red water jug and gradually regained composure, only sputtering and wheezing once in awhile.

Frightened, I told him that I would take him home.

Dad refused. "I'm fine," he said.

"You are not fine," I replied. "We have to get you some help."

Dad had suffered from coughing spells before. He told me later about coughing fits he'd had while driving the tractor, literally blacking out for a few seconds and coming around enough to keep the tractor on the row. Years of hog and grain dust were obviously taking a toll, but this was the first time I realized the seriousness of my Dad's bronchial asthma. After several doctor visits and an extended hospital stay by Dad confirmed my worst fears, I knew that nothing would be the same for my father, our farm, and me ever again after that day.

I had graduated from college a year earlier, but college didn't prepare me to assume all the financial responsibility of the farm quite yet. I wasn't ready mentally, emotionally, psychologically or spiritually. I had always wanted to farm and had planned on farming after college, but not on my own right away, and certainly not without Dad.

But his doctors were clear. He had to quit farming immediately. So almost overnight, it became my turn to carry on the family tradition. I invoked the help of St. Isidore many times as I drove the tractor or combine up and down long rows of crops that first year after the diagnosis, worrying that I wouldn't be up to the job.

After I had completed the first harvest that I'd worked on my own—from planting to combining the grain—I jumped down from the combine steps and knelt on the cold ground, humbly thanking God for helping me along the way. I also thanked St. Isidore for sharing his white oxen and their angel drivers now and then to assist me.

Today, Dad still comes to the farm nearly every day, because farming is in his blood. He has learned to live with his asthma, taking the proper medication and staying away from hog and grain dust, doing things that don't irritate his condition. He has since suffered through prostate cancer, shoulder and back surgeries, and a number of other maladies. He continues to be the most devout, prayer-minded person in our immediate family. I can say with certainty that a rosary is being recited for my family and our family farm several times a week by my parents, who pray together every morning, carrying on the lovely rosary tradition of my Grandpa and Grandma Arens.

Isidore died around 1130. King Alfonso of Castile credited a vision of Isidore for showing him a secret path by which to overtake and conquer an invading Moorish army in 1211. The relics of Isidore were credited with curing King Philip II of Spain of a mortal fever in the late 1500s when they were brought into his bedroom. And I credit Isidore with aiding me through a challenging farm transition period.

Although he was revered by his countrymen for centuries, Isidore wasn't officially canonized until 1622, at the urging of the royal Spanish family. He and wife Maria are now venerated as the patron saints of Madrid, Spain. Their feast day is celebrated in the United States on May 15, midway through planting season. They also have become the patron saints of American farmers, farm couples and families, farm workers, and others living on the land. They've also been adopted as the patron saints of the National Catholic Rural Life Conference.

Around Crofton, I see a few quiet folks like my parents, who often touch the lives of my family but never will be venerated outside their own families and friends. Their old bones will not be considered relics. Yet, they go through their daily lives like still water—calming where there is chaos, soothing where there is grief, and providing quiet service for others. The Church will never canonize these people, and miracles will not be attributed to their intercession. Yet, I admire them and they inspire me the same way the stories of saints inspire millions of the faithful. We need not look to the one-dimensional "heroes" of television or movies or sports but only to our own families and the saints held up to us by the Church to find real heroes of depth and substance.

Married with Children

I had given up on love. I had been "dumped" so many times that I had resigned myself to the fact I could be quite happy as a bachelor farmer.

Still single, I considered other options I hadn't thought about since third grade at St. Rose School. One of my priest friends, after hearing me eulogize a friend at an extremely sad funeral Mass after a tragic death, asked me to seriously reconsider the priesthood as my vocation. This thought had crossed my mind already. I was getting older and more "set in my ways." I had graduated college and was farming full time, taking over the reins from Dad. But entering the seminary would change not only my life but also the future of our family farm.

I didn't believe that my being single at that point in my life was reason enough to think becoming a priest was my predestined vocation. In my mind, you have to truly feel the calling to remain celibate because of a commitment to the priestly vocation, not the other way around. I had no such epiphany. Still, I agonized over the idea for several nights after the priest mentioned it. Was his gentle nudging God's way of calling me to ordination?

I thought of how God called Samuel several times at night while he was asleep before Samuel finally recognized God's voice. So, for one of the first times in my life, I actually followed the prayer I had learned as a teenager from youth faith-formation weekends: "Let go. Let God." I told the Lord, "Whatever you want is OK with me." And I meant it.

About the same time, a good friend asked me if I would attend a Cursillo weekend. I had participated in similar youth-oriented retreats—called Search weekends—in high school. I was familiar with the Cursillo movement, because of my parents' involvement, so I agreed to attend one of the programs for men. Sitting in a room at a Cursillo weekend in Immaculata Convent in nearby Norfolk, I was astounded at what I heard from the speakers. Husky, mature men were speaking about their faith in ways I had never heard before. It was as if I had found myself among aliens—guys who minutes earlier had been talking about football were now

> *Guys who minutes earlier had been talking about football were now sharing their deepest feelings about their own spiritual lives.*

sharing their deepest feelings about their own spiritual lives. I listened intently as each speaker related in eloquent, emotional terms, his love for his wife and children. The experience moved me in ways I wasn't prepared for. I realized that I wanted more than anything else to experience such deep devotion to a family of my own. I coveted *that* vocation for *my* life. It was an epiphany of a different kind: I was not called to be a priest; I was called to be a husband, a father…and a farmer.

My wife-to-be, Donna, and I went on our first date a month after I attended Cursillo. To this day, I do not believe it was a coincidence.

After we were married, Donna finished the school year at St. Leonard School in Madison and accepted the kindergarten teaching position at St. Ludger, an elementary school in Creighton, Nebraska, thirty-five miles from our farm. She didn't look forward to the long daily commute, especially during the cold and snowy winter months, but she loved the job.

We discovered Donna was pregnant with Lauren around our first wedding anniversary. Donna struggled through terrible morning sickness, to the point that she was often ill driving to work and even during the morning children's Mass at St. Ludger. She lost eleven pounds before she gained a single pound during the pregnancy. Fortunately, our family doctor, C.J. Vlach, was a kind, humorous fellow with a tender heart and a fatherly way. He guided us, as newly-expectant parents, through the pregnancy experience. He was a calming factor for us when we were weary of hearing our friends impart their tragic miscarriage stories. Finally, he told us during one visit, "You both just have to put the whole thing in God's hands and stop worrying so much."

He was the perfect physician for us. So, when Donna had to have an emergency Cesarean-section after pushing for several hours before Lauren was born, we appreciated having Dr. Vlach in the operating room.

Both Donna and I cried when we first heard Lauren burst onto the scene, wailing at the top of her lungs. I carried her into another room where the nurses cleaned her and warmed her tiny body. Dad, Mom, Donna's mother, Jean, her sister, Rhonda, and my great-aunt, Sr. Blanche—a long-time Benedictine nun and retired but legendary OB nurse—were waiting to see Lauren. Sister Blanche immediately looked her over from head to toe and with a reassuring smile that reminded me of Grandma Bickett told me our new baby girl was "perfect."

When Donna was pregnant with Taylor two years later, Dr. Vlach was quite ill, fighting cancer. Although he was weak and visibly drained, he continued to guide us through the pregnancy as Taylor's birth drew near. When the day came for a planned C-section, although officially he was on medical leave, Dr. Vlach joined us in the delivery room to witness Taylor's birth. He cried with joy and hugged us when our second little girl was born.

By the time our son, Zachary, was born, both Sr. Blanche and Dr. Vlach had died. Zachary, though, was fortunate to have two special angels watching over his birth. As I went through the ritual of showing our little boy through the nursery windows to waiting relatives, I glanced at the wall near the nursery to see a framed memorial photo of Sister Blanche smiling back at me, and as I visited the hospital chapel, I noticed a new memorial portrait of Dr. Vlach.

When Taylor was baptized, contractors were in the middle of a major remodeling project in the fifty-year-old St. Rose Church, so weekend Mass was typically held in the church hall. We wanted Taylor baptized in a church, so our pastor at the time, Father Bob, baptized her in what was then St. Rose's mission parish, St. Joseph's Church at Constance, which happened to be the church where my paternal grandparents had been married. The historic, country church served as the perfect back-drop for our family to renew the sacred ceremony of adding another Arens to the Catholic fold.

Another priest, Father Joseph, had come to our parish during the time between the births of Taylor and Zachary. After Zachary was born, Father came to the farm one night to administer baptism classes for Donna and me and one of Zac's godparents. Father quizzed us on basic matters of the church, things I knew but wasn't able to verbalize as clearly as when I was a fifth-grader at St. Rose School.

Father Joseph's baptism class reminded me of parents' continual need to remain fresh and renewed in understanding the teachings of Jesus and the basic doctrines and practices of the Catholic Church. He reminded both Donna and me of the awesome responsibility we had agreed to on our wedding day—to teach our children about Jesus and to foster a Catholic identity in our household.

Once we had three children, Donna and I could no longer "divide and conquer" to enforce our daily household routine, because the kids outnumbered us. The challenges of parenting make me wonder how my grandparents had raised eleven children. A friend advised me, using a football metaphor: "You just have to start playing zone defense, instead of man-to-man."

Early on in our married life, Donna and I enjoyed going to movies, attending ballgames in town, and spending time together in the pasture fixing fence or riding in the tractor. Donna tagged along with a friend and me as we played disc jockey for several weddings and anniversaries. With an arsenal of hundreds of popular compact discs and tapes in our personal collections and an old but quite sufficient sound system at our disposal, I found it rewarding to choose the right song at the correct time to lure everyone out on the dance floor.

Nearly every song we cued on the CD player brought out a dancing crowd.

One time we played music for a neighbor's wedding dance in Crofton's auditorium. It was a festive celebration from the playing of the first song when the happy couple graced the floor. Nearly every song we cued on the CD player brought out a dancing crowd. When we played the Chicken Dance, the Hokie Pokie, the Bunny Hop, and a local favorite, the Flying Dutchman, the auditorium floor was so crowded that nearly all of the booths and tables were empty.

My neighbors and I look at celebrations like this as opportunities to "kick up our heels." I can't help but think of Jesus attending the wedding feast at Cana: At the urging of his mother he performed his first miracle by changing the water into wine, thereby saving his friends' party and revealing the love of food, drink and fellowship he was known for throughout his life.

Today I no longer do the D.J. bit, and Donna's and my ideas of fun have become much more kid-oriented.

It is true that kids say and do the most entertaining and embarrassing things, at least in the eyes of their parents. During a Christmas Eve Mass, when Lauren was about three years old, Father Bob asked all of the children in church to sit before the altar so he could read a special Christmas story to them.

I took three-year-old Lauren's hand and led her along the outside aisle toward the front of the altar where Father was already sitting and children were gathering, facing him. I whispered to Lauren, "Go sit by Father." She took my meaning literally. While the other kids faced Father, she plopped down beside him, leaning on his knee. I stood along the aisle, waiting to see what she would do. Father began reading a lovely childrens' Scriptural rendition of the Christmas story. About the time the innkeeper sent Mary and Joseph to the stable for the night in the story, Lauren grasped Father's vestments gently and whispered in his ear, "I have to go potty."

Father smiled and replied quickly, without hesitation, "Go right ahead." Then he began reading again.

Lauren trotted toward me so I could take her to the bathroom, but she stopped short of where I was standing. Spinning around and holding her hand up in the air, she said rather loudly to the priest: "But I'll be right back." Everyone in the front pews saw what had happened and burst into laughter.

One early morning, I was driving Lauren and Taylor to the babysitter in town for the day because Donna was taking summer courses and would be gone. I had alfalfa hay in the field ready to bale, so I poured Cheerios in a bowl and allowed both of the girls to munch the second course of their breakfast on the road—a practice I detest but one that was necessary on that particular day.

As we drove down our long driveway, the adjacent oats field was golden, nearly ready to harvest. I watched the tall oats ripple in the wind, like waves of water in the ocean. Lauren tugged on my shirt.

"Daddy, where do Cheerios come from?" she said curiously.

I nearly made what would have been an unforgivable mistake for a farmer by saying, "From the grocery store." Too many children think that milk comes from a carton, hamburger from a fast-food joint, and fruit from a can. Fortunately, I caught myself and looked upon the oats field again. "Lauren, your Cheerios and a bunch of the cereals and breads you eat every day start right here on the farm, out in a field like that, because oats is one of the grains used to make Cheerios."

It was a valuable teaching moment. I took advantage of the opportunity to talk with my little girls about the miracle of a seed germinating in the earth, growing up with sunshine and rain from heaven into a crop that we harvest and process into food for humans and livestock. "You know how we pray during the Lord's Prayer, 'Give us this day our daily bread'?" I continued. "That is how we ask God to help us grow our food in the fields."

It saddens me that we Americans value food so little. About a third of us don't even stop to eat properly, rushing through fast-food restaurant drive-through lanes on the way here or there, shoveling food into our mouths without taking time to enjoy it or the company of our family and friends. I want my children—farm kids—to appreciate the miracle of food and consider it a genuine gift from God.

Donna's mother, Jean, marvels at how Donna and I could produce our two daughters, who are of entirely different dispositions. When Lauren was a baby, we probably jumped up at every squeal she made in the middle of the night, and consequently she was a restless sleeper from the beginning. Many nights, I held her in my arms as she wailed for no apparent reason. I tried everything I could to calm her. We tried letting her "cry it out" and we tried holding her in our arms, and every known technique between the two extremes. While holding her in the middle of the night, I was often exhausted and at the end of my rope. I tried singing her to sleep with TV show tunes like theme songs from the "Beverly Hillbillies" or "Brady Bunch." One of the only songs that would calm her was "Hail Mary, Gentle Woman." Knowing how poor my singing voice is, I'm surprised my singing didn't make her cry louder.

Taylor, on the other hand, slept through the night from the time she was three months old. We would often creep into her room to make sure she was OK, only to find her sleeping soundly.

Lauren and Taylor love farm life. Lauren enjoys being a Clover Kid in our local 4-H club, but Taylor might be the child in our family who truly understands what it takes to be a farmer. This past summer, I awoke one morning to the sound of hogs on the loose, snorting and digging in Donna's flowerbeds. I ran to the barn to fix the fence they had broken and lure them back into their pen. Taylor charged from the house wearing her pajamas and sandals, ready to assist. "Just stand there by the machine shed and yell at the hogs when they come near," I told her while frantically trying to round up 60-head of hogs, with several romping down our driveway and others milling around inside the shed. I prayed for help from the Good Shepherd, hoping he might be a decent hog man too. Just then a group of ten hogs ran from the shed, stopping in their tracks to look at Taylor. "Get back in the pen, you pigs," Taylor yelled with authority, raising her arms above her head. They snorted and turned toward the barn's open door, back into the pen where they belonged. Taylor and I spent two hours luring the rest of the free-roaming swine safely into their pen. After we had successfully accomplished our mission, we walked to the house together,

> "Get back in the pen, you pigs," Taylor yelled with authority, raising her arms above her head.

sweaty and smelly, but satisfied we had corralled all of the escapees.

It is too soon to predict what type of personality Zachary will bring to our family. He already has charmed nearly everyone he meets. He was only three weeks old when Donna, still on maternity leave, took him to school during the Catholic Schools Week open house so her students could meet him for the first time. At one point, a number of eighth-grade girls were holding him gently. He was sleeping soundly. Observing this scene, Sister Charlotte noted that Zachary needed to learn that if girls are paying attention to him like that he should at least stay awake! I'll have to remind him of that when he is a teenager.

Yes, every child is precious. I am amazed daily at the brilliance of God's creation, and I'm grateful for the unique three children we have been given.

Donna's and my life's priority list—if we had to boil it down to three things—is faith, family and farm. We try to keep these priorities in perspective, but it is tough. I take my farming responsibility seriously, for example, yet on some evenings when my neighbors are planting corn, I might be inside the house helping bathe the kids so Donna doesn't have to stay up until midnight to finish grading the stack of papers she brought home from school. I try to be there nearly every night to escort my children to bed, tuck them in, and recite their nightly prayers with them.

On the other hand, when alfalfa needs to be baled or during planting and harvest time when storm clouds are threatening, Donna sacrifices time she might need to complete her schoolwork in order to take care of the kids so I can work in the fields late into the night.

It is a balancing act. When I'm in the fields, I feel guilty about not spending enough time with my family. When I'm home, playing with my children, I feel guilty for not working hard enough. We try to utilize our priority list as a barometer in making these decisions, but it causes intense discussions between Donna and me.

We both know, however, that if we want to abide by the priorities we have established we need to keep our time for God at the top of the list. I often think of how Jesus reacted to the old woman who dropped her last pennies in the church collection basket. He told his disciples that the woman had given more than many of the wealthy. When they questioned him further, he explained by saying the woman gave her last pennies—money she needed to live—while the others gave of their surplus—funds they didn't really need. I think true faith and stewardship go hand in hand, not only in supporting the land but also in maintaining a rural parish and school.

Helping our children understand these priorities and take them to heart is difficult, but it is worth pursuing, because learning to set priorities and stick to them will serve them well as they grow old. In this case, I suppose, talk is cheap and actions speak louder than words.

Come with Me into the Fields

You'll do just fine," Dad said as he jumped into the pickup. I watched him drive out of the field from my perch in the driver's seat of our Farmall 560 tractor.

I had limited tractor-driving experience. Months earlier, I drove our little Farmall Super H in low, "granny" gear at a whopping speed of 1 m.p.h. through a myriad of small square alfalfa bales, as Dad and our hired man, Robert Schaefer, loaded the hefty bales on the wagon I was pulling. With a full-time hired man like Robert on duty, most of the tractor driving when I was a teenager on our farm was up to him. In the world of steering a tractor through obstacles like field borders, creeks and fences, except for my brief stint on the Super H, I was a complete novice. However, on this particular day, it finally was my turn to contribute to the workload in the field.

I had bounced around the rear of our seeder-wagon all day long, scooping oats seed into the hopper of the seeder that planted the field. Now, it was time to pull a drag through the field to incorporate the seed into the fresh spring soil, to smoothen the clumps of soil for a firm seedbed.

I stepped on the clutch of the 560 gingerly, grinding gears as it slipped into fourth gear. "If you can't find 'em, grind 'em," I could hear Robert teasing me in my mind. I pushed the throttle forward and stepped off the clutch. The tractor lunged forward, pulling the drag diagonally across the field. I came to several shorter point-rows, where the field isn't squared off, making it more difficult to maneuver the equipment. West Bow Creek meanders through our farm along the north side of the field in a disorganized manner, creating bends and loops along the fenceline and field border. I was wary of its steep twenty-foot banks.

I vividly recalled my preschool days, coming home with Mom from a Home Extension club meeting at a neighbor's farm and seeing Dad washing up at the kitchen sink, with cuts on his head and dirt covering his clothes.

"What happened?" Mom asked.

"The tractor rolled into the creek," Dad replied.

Mom laughed. Dad didn't.

He had driven too close to a creek bank and the tractor and wagon he was pulling tumbled down the steep slope. He jumped off and ducked into a heavy metal culvert protruding out of the creek bank. The wagon hit the corner of the culvert before it splashed into the water, with the tractor, landing upright, directly behind it. His life had been spared only by the grace of God.

I thanked God for saving me from what would have been a grisly accident.

With that memory clear in my mind, I made the first turn with the drag along the creek. As I came out of the turn, a cable holding the drag in line started to climb the rear tires of the tractor. I had turned too sharply and the rear tire was picking up the cable and pulling the drag on top of me. Something made me push in the clutch pedal. Everything halted. Breathless, it was as if my heart had stopped. The tractor was still running full-throttle. The cable was wrapped up on the rear tire and the drag section with its sharp spikes had started to climb in the air, hovering above me in the open tractor seat. I sat stunned in my seat for five minutes, my heart pounding so loudly I was sure neighbors could hear it. Finally, I realized that if I shifted the tractor into reverse the cable and drag section would fall back to the ground.

After escaping the predicament, I looked behind me. The equipment was in place again. I thanked God for saving me from what would have been a grisly accident. It was the first time I really understood that our good Lord is with me when I'm in the field. I felt as if he had been sitting on the tractor wheel-fender beside me, stepping on the clutch pedal for me, giving me the common sense I needed to manage the situation. I shifted the tractor into gear again, with a new appreciation for the spirituality of tractor-driving. I finished dragging the field, and have never made the mistake of turning too sharply with a drag again.

Hundreds of farmers die every year in accidents with machinery and livestock. Farming ranks regularly as one of the top three most dangerous occupations in the country. Nearly every farmer can recount a litany of close calls, brushes with injury, or even near death experiences. I have had several potentially life-threatening situations myself.

Years after this incident, I was trying to put an identification tag in the ear of a newborn calf when the protective mother cow came from behind me, lifted me into the air with her head, smashed my body against a metal bale feeder. She continued to charge until I finally was able to crawl inside the feeder for safety.

Another time I was combining corn one dark night in a field with power poles and lines running diagonally across the crop rows, I glanced out of the cab window beside me in time to notice one of the poles just missing the side of the combine by a whisker. It was too late to turn out of my position, so I waited for something to crash alongside the huge machine. But it never did. I continued safely down the row.

I can't tell you how often I've been exhausted in the tractor seat, nearly falling asleep at the steering wheel, when I felt a start, a bolt, or a nudge awaken me, just in time to turn away from a dangerous obstacle. I know I'm blessed with someone watching over me in the fields.

I believe God uses the people in our lives as messengers and helpers, even if they don't know it. When Dad was recovering from an accident, where he lost his ring finger when his wedding ring caught on a bolt as he swung out from the truck box, our neighbors pitched in without being asked, helping haul cattle scheduled for market.

When Dad was hospitalized the first time because of asthma, Ron Tramp, our neighbor across the creek, spent hours at our farm, helping me sort hogs for market and haul calves from the pastures. At the time, I was dazed and uncertain of myself because of how rapidly my farm responsibilities had increased. Ron served as a mentor, calming me down so I could complete the work before me. Without my long conversations with him about farming and life on the farm, I perhaps, would not have made the transition so easily.

Dale and Charlotte Arens and another neighbor, Gary McShannon, live over the hill from our farm, serving as angels for Donna and me. Gary has pitched in during emergency situations, helping me on one occasion to change the position of a cow having a calf on a muddy pasture slope. Not an easy task. Dale once risked his own safety on a cold, snowy morning, to ward off an angry sow while I moved her litter of pigs into the warmth of the farrowing house. I especially recall the evening of the September 11 terrorist attacks, when Dale drove to our house on his ATV. Donna and I were inside, glued to our television, watching reports from New York and the Pentagon. "Your cows are running down the road up west," he said. "I'll help you get them in."

I jumped in the pickup and we quickly drove to our west pasture a mile away, coaxing the cows inside the fence. As we stood in the dusk fixing the barbed wire fence the cows had broken, I looked up in the sky where we would normally see multiple airplane smoke trails streaking across the horizon. On that night, when all of the flights were grounded, we both felt eerie about the calm, clear sky, and we shared our mutual anxiety about the future of our country under attack.

Dad has always said God will not give us more to endure than we can handle. Sometimes a friend, neighbor, stranger, parent or grandparent offering a helping hand at the right time is the lift we need to keep going. Acknowledging that God is present to us in our daily lives through others is an appreciation that God plays an irreplaceable role in our daily lives.

One of my favorite songs we often sing during Mass goes: "The harvest is plenty. Laborers are few. Come with me into the fields." In the song, God is not speaking of grain we harvest but of harvesting souls for Christ. Many Catholics are uncomfortable evangelizing in an outspoken manner. For a reserved, German farmer like me, witnessing my faith to others is quite difficult. I was taught to express my faith within the safe confines of my immediate family and our parish setting. Outwardly discussing Christ is somewhat foreign.

We witness faith in ways that aren't so apparent, often through kind actions toward others.

However, we farmers easily acknowledge God's presence. There is no greater faith than placing a tiny seed in the soil, counting on the miracle of life to germinate and produce grain so that the farm family can make a living.

We witness faith in ways that aren't so apparent, often through kind actions toward others. That form of faith expression allows us to answer in the words of another of my favorite Mass songs, based on Isaiah 6: "Here I am Lord. Is it I Lord? I have heard you calling in the night. I will go Lord, if you lead me. I will hold your people in my heart."

Rural people have an innate ability to heed that call, acting out their faith through their willingness to "love thy neighbor as thyself." Small town residents know each other—bumps, bruises and all.

When my teenaged nephew from Omaha stayed with us for two weeks one summer, I was worried that he would miss the bright lights of his urban home. Over the time he was on the farm, he learned to steer a tractor and combine, drive a pickup in the pasture, chase cows, fix fence, and appreciate a starry night sky. He could ride a bicycle around the farm, with no worries about intrusion from strangers.

When I waved at several passing pickups while we were driving to town running errands, he asked, "Do you know all of those people you are waving to?"

I replied, "Yeah, I know most of them. I wave to the rest, just to be friendly."

In town, I introduced him to acquaintances who work at the feed store, grain elevator, bank, and grocery store. "People are so friendly here," he told me as we drove back to the farm. "Everybody seems to know one another."

The cliché about small town life that everyone knows your business is true, for the most part. There aren't many secrets when your community is close-knit. However, if tragedy strikes your family, it is comforting that you can count on people to support you, assist you, pray with you, and stand up for you.

I listened to Brother David Andrews from the National Catholic Rural Life Conference speak about ethics in agriculture when he spoke one time at Mount Marty College in Yankton. Brother Andrews reminded those in the audience that being a good steward is helping carry out God's plan for the Earth and for humanity.

A farmer friend explained a similar theory to me. "The best way I can be a good steward of the land is for me to leave this land better than I found it," he said. "When I'm long gone, hundreds of years from now, I hope there is no trace or sign that I've ever been here."

I've learned to pray for guidance from the Holy Spirit, as tough ethical questions in farming have made it more difficult to tread lightly on this planet. Am I using all of the cultural tools and practices I can to raise my crops in ways that are friendly to the environment? Am I taking measures to insure the overall husbandry and comfort of the livestock in my care? Am I teaching my children how to be good stewards, to care about water and air quality? Am I being a good neighbor to the folks who have helped me many times? These are questions I ponder in nearly every decision I make on the farm.

The United States Conference of Catholic Bishops has developed an interpretation of basic Catholic social teachings that apply directly to the daily management decisions farmers make. In the "Catholic Social Teaching and Agriculture" section of their insightful 2004 report, *For I Was Hungry and You Gave Me Food: Catholic Reflections on Food, Farmers and Farmworkers*, the bishops called upon farmers and rural policymakers to consider several principles in their decision-making, asking them to use the Scriptures as a guide for bringing food to the dinner tables of the world. The right to food is part of every human's right to life and dignity, so it is our responsibility to work to provide food for every human being, the bishops wrote. "The human person is not only sacred but also social," they said. "Rural communities and cultures, with their focus on family life, community, and close ties to the land, serve as welcome signs of these social dimensions of Catholic teaching."

In the report, they addressed the way American farm policy might threaten farmers in other lands. The bishops believe the world's poor and vulnerable should be protected. They wrote, "An important moral measure of the global agricultural system is how its weakest participants are treated."

Dignity of farm workers and laborers and their rights to a healthy workplace, economic freedom, fair wages, and access to health care and time off should be defended at all times, said the bishops.

The U.S. bishops discussed the importance of feeling solidarity with the world's people: "Agriculture today is a global reality in a world that is not just a market. It is the home of one human family."

Finally, the bishops reminded us creation is a gift from God. It is not something we should manipulate to our private economic benefit. "All of us, especially those closest to the land, are called to special reverence and respect for God's creation," they explained.

They laid responsibility for stewardship beyond the boundaries of the farmyard. "While rural communities are uniquely dependent on land, water, and weather, stewardship is a responsibility of our entire society."

I'm sure God is active in guiding farmers in these important ethical land stewardship questions. In this time of rapid, industrial changes in agriculture, the Lord's guidance is needed more than ever. We can't make the world a better place by ourselves. We need a community to do so. It seems to me that God is calling my neighbors and me to roll up our sleeves and promote positive changes in our little corner of creation.

Family Tradition

My neck started to ache slightly, and I was getting a throbbing headache. As I looked up at the ceiling, a glob of wet dirt plopped down on my forehead. As I brought my head back down, water and dirt ran down my nose, into my eyes, and down the front of my already-soaked tee-shirt. I wiped the mud away from my eyes. I was soaked and grimy from my blue Minnesota Twins baseball cap down to my brown leather farmer's boots. The power washer hummed in the background, and water continued to drip from the rafters of the barn I was cleaning. Almost ninety years of grime—including horse manure, cow manure, hog manure and fly specks—were washing down onto the rough concrete floor beneath me. I thought about the heritage of this old barn as I continued my dirty job.

I was preparing for the first Arens family reunion. Over 200 of my relatives were planning on a family reunion here at our family farm. My brother, Paul, and I had cooked up the idea when we were washing one room of the barn for use as a seed storage area. Just brainstorming, as we often do when we're together, we thought about what a kick it would be to clean the entire barn and invite all of our living aunts and uncles and cousins and kin back to the farm that many would recall from their youth and some may have never even seen.

Grandpa and Grandma Arens had been gone for years, but Paul and

I wondered aloud if the lure of the home place was still alive in our extended family. It was a crazy idea that would require months of preparation, and there were no guarantees that anyone would attend. Paul and his family live two hours away, but he excitedly offered his expertise in the landscaping business to spruce up the land around the farmstead.

I wondered what might become of our farm a hundred years from now.

Now, here I was, just a few days before the reunion, scrambling to get the barn clean and ready, because it seemed that nearly everyone was coming "home." The connection of our family to this farmstead turned out to be downright mystical. The branches of our family tree have sprawled across the world, but the roots remain in the shadow of Grandpa's barn. I was to learn firsthand that nowhere are familial relationships and history and sense of place more revered than on a family farm.

Many farms like ours have been abandoned and torn down over the past few decades. As I continued to wash the barn, I wondered what might become of our farm a hundred years from now. I recalled the words of the great Nebraska orator and three-time Democratic Presidential candidate, William Jennings Bryan: "Burn down our cities and leave our farm land and the cities will rise up again like magic. But burn our farms and grass will grow up in the streets of every city in America."

Vacant farmsteads around the countryside are the last reminders of families who took a chance on the land and lost. Many others became successful farmers but had no one to pass their legacy on to. Usually it's the barns of abandoned farms that most easily catch your attention and imagination. Often they are worn and weathered old specimens with holes in the rooftop. Sometimes there is an empty farmhouse where the occupying family spent their lives, now shrouded in overgrown trees and grass that once were kept dutifully trimmed and mowed.

What we don't notice as easily are the ghosts of farmsteads that have disappeared completely from the landscape. We can't see the enormous cornfield that was once at the center of a concrete shopping mall parking lot. We forget that a family once carved out a living raising crops, with barns housing chickens, pigs, lambs and an old milk cow named Betsy, where tracts of houses now stand.

Uncle Ab and Aunt Ceil's farmstead along West Bow Creek, just across the field from our place, is one of those ghostly farms. I grew up there, too, playing around the place with my cousins. I knew their farm, much as I know our home place. Even after Ab and Ceil were gone and the farmstead abandoned, I appreciated that their buildings were still there. I didn't realize how much I appreciated the physical presence of that farmstead until the day those buildings disappeared.

One morning after I rolled out of bed, Dad immediately announced, "Curt, we've got a job to do this morning."

I had noticed bulldozers parked on Ab and Ceil's place for a day or two. Now it was time for the dozers to do their thing. Dad said the new owners were planning to demolish the place and plant crops where the abandoned buildings once stood.

That wasn't unusual. We had torn down an old house and barn on another farm that we'd rented years ago, just to make room for cattle-sorting pens. But the news of the demise of Ab's place hit me like a load of bricks.

"The house and barn on Ab's are going down today," Dad said. "They're gonna light the fire this morning, but we can have anything we think is valuable in the house if we get it out right away."

Our family rented and farmed 160 acres of the land surrounding Ab's place after Ab retired from farming. After Ab and Ceil died, our family sublet the house to several different families over the years, while we still leased and operated the farmland. Even while other families lived there, occupying the house, garage and barns, no one could call the farm Kleinschmit's, Shoat's or Ibach's. To the neighbors, it was always "Ab and Ceil's" place.

When the owners decided to sell the land, however, we couldn't afford to buy it. The renters left. The buildings began to deteriorate. The farm was sold again, and the new owners decided to bring the buildings down.

I fed hogs and did a few other chores while Dad hastily gathered some tools. We drove the pickup into the driveway of Ab's place and around in front of a majestic cottonwood tree that had always looked like a giant to me as a youth. I hadn't been on the farmstead for four years, although I drove by it several times a day. But I hadn't paid attention to its condition. I hadn't noticed broken windows and weeds growing six feet tall around the porch. I hadn't noticed how Ceil's sprawling willow tree had been reduced to a few weak sticks. Her tulips remained, ready to burst in spring colors, but the giant cottonwood was no longer the stately monument I remembered it to be. The barn had been altered and remodeled several times, so it didn't even resemble the building where Ab and Ceil had raised broiler chickens each summer.

Dad and I didn't have to open the porch door to enter the house; it had been torn off. Fly strips hung from the ceiling of the kitchen. I remembered how warm that kitchen had been whenever we visited. Ceil was always baking, and my brother and I looked forward to hot cookies from her oven that burned our mouths. They somehow tasted better that way. The sweet smells of the kitchen were now gone, giving way to the odors of mice and dust.

The bulldozer operator burst into the house and said he was ready to start as soon as we were finished. He hovered over Dad and me like a vulture waiting for a meal. There wasn't much time. I began cutting wires and unscrewing lag bolts from an electrical box Dad had put in the house for the first renters after Ab and Ceil died. It was still in good shape, so we tried to salvage it. I smashed my hand against the wall and my knuckles began to bleed. I used my pliers to unfasten the lags in the back of the box and pinched my thumb between the arms of the pliers as it slipped from the head of a lag. I cussed at the box and the pliers, the house and the situation. I didn't like being rushed into saying goodbye to Ab's place.

I asked myself if he had the peculiar sense of loss I was feeling.

Dad hurried to the basement to see if the hot water heater was worth salvaging. He sprung up the stairs to check on my progress. I asked myself if he had the peculiar sense of loss I was feeling. All he seemed to care about was getting that hot water heater or electrical box and getting out.

Finally, the wires were all cut and the lags loosened. We tried to release the box from the wall, but it was stubborn. Dad used a pry bar to pull the wood paneling away from the wall. Large chunks of the finish flew from the walls. The demolition seemed violent.

We tore at that wall to finally release the electrical box. The hot water heater was rusted, so we removed two old pressure tanks and an outlet. That was all we could find to salvage. The cupboards were gone. The crucifix on the wall was gone. The photos were gone. The chairs, tables and even the bathtub were all gone. A few ragged curtains and memories remained. There was a faded spot on the wood paneling where perhaps one of Ab's fishing plaques had hung. You know the kind: "Old fishermen never die, they just smell that way."

Dad and I walked silently through the house. Looking into the living room, I could see an image of a pale, emaciated Ab lying in a hospital bed and Ceil standing beside his bed, both receiving the Eucharist from Dad on a Saturday evening.

We walked into their bedroom. The windows were smashed. Dad still said nothing. We walked outside. The dozer man said he was ready to start. "When do you think this old house was built?" he asked. "Probably hundred years old, don't you think?"

"I don't know," Dad replied. "The basement is made of rock and mortar, so it's pretty old."

The dozer man observed, "Look at those peonies coming up. They must have liked flowers with all the different stuff growing around here. A lot of these old places are gone now. Everybody needs more acres to make a living, but there just aren't many farmers left in the country who need houses."

I was sick to my stomach. The dozer man had already pushed many of the trees into the side of the house, including the old willow. Dad climbed into the pickup beside me. We began to drive away. We both looked back as the dozer man started the fire. Suddenly, flames broke through the windows and the siding. The building was engulfed, smoke billowing into a bright, blue sky. Dad raised his hand and waved goodbye to the old farmstead. He turned to me, moisture welling in his eyes.

I still see shadows of Ab's place when I drive by almost daily. We can look out our living room window and see the cedar trees Ab planted along the highway outlining their old place, but crops now grow where the house and barn once stood.

One day a couple of years ago, I noticed as I drove by a single peony flower blooming above the soybeans where Ceil's garden used to be. She had planted that flower long ago, and now it bloomed almost defiantly.

That peony reminded me of the lessons of faith, hope and love from 1 Corinthians 13 that I had learned firsthand from Ab and Ceil. Because of those lessons, launched by vivid memories of the way a loving wife cared for her dying husband, I now pass Ab's place with a new perspective. The buildings may be gone, but that special place they created in my life will never fade.

When I was young, each spring it was a tradition in our family to burn the dried palms from Palm Sunday of the previous year and drive as a family to all of our fields to spread the ashes. In the same way bishop Mamertus asked for God's protection centuries ago, we too prayed for a successful crop that year and for guidance and safety for our farm family.

But I had never heard of "rogation" until I read an article in my local newspaper announcing the three pastors of Crofton's churches, including our pastor, Father Joseph, were planning a rogation or blessing of the seeds. Dad recalled Rogation Days at church from his youth, but the practice in this area had been basically phased out.

The idea of asking for blessings and protection of farms, fields, parishes and villages began when the fifth-century French bishop, Mamertus, prayed to God, asking for protection from drought and pestilence on the countryside.

In Dad's youth, a major rogation occurred each April 25, when farmers and gardeners brought their seeds to Mass. During the Rogation Mass, everyone processed around the parish grounds, asking God to bless church and school buildings and the boundaries of the parish or even the town. During the procession, parishioners recited a litany of the saints. Other minor Rogation Days were celebrated on the three days before the Feast of the Ascension.

Although Crofton's modern-day rogation was not the same liturgically as the rogations of my Dad's youth, Mamertus would have recognized several of the Scripture readings and blessings. At one point in the service, everyone brought forward a sample of the seeds they would be planting in their fields and gardens that spring. Each of the pastors of Crofton's churches took turns blessing the seeds, asking blessings on our soil, farm implements, vehicles and livestock.

I brought a small bag of seeds that Lauren had collected for a class in school, representing our 600 acres of farmland. Father Joseph asked a simple blessing on everyone's seeds, sprinkling the little bags and small pails with holy water.

The rogation was an important reminder to me that all things we enjoy on our farm and in our rural community come from God. We have a partnership with the Divine, a kind of cooperation that has been passed on from generation to generation of farmers. That type of partnership carries with it some key spiritual responsibilities, however, and I am continually honored by the opportunity to be one of the members of my family to carry on that legacy.

It took a week for the barn to dry out after it was power washed, but the family reunion day dawned sunny, hot and windy. My cousins had helped erect a large tent outside the barn the night before for extra seating. We had a flatbed trailer as a stage for the family program. Around noon, family members started arriving. Everyone parked in a pasture a few hundred yards west of the farmstead, so Dad and several of the uncles and cousins chauffeured guests back and forth on golf carts.

The next day, everyone attended Sunday Mass together at St. Rose, in memory of my grandparents and all our deceased family members.

Displays, a family video, old albums, and my grandparents' clothes and photos decorated the barn I had cleaned only a week earlier. Reminiscent of the old Sunday gatherings my grandparents hosted so often at their farm, cousins played horseshoes, sat in lawn chairs in the shade and visited, and played ball and other games with their children and grandchildren. We even prepared the back part of the barn for a dance later that night.

All of Ab and Ceil's children, those cousins I knew so well growing up, returned to Grandpa and Grandma's farm, gathering together with their extended family for the first time in twenty-five years.

The next day, everyone attended Sunday Mass together at St. Rose, in memory of my grandparents and all our deceased family members. We then visited the cemeteries of St. Rose and St. Joseph parishes, laying wreaths on the graves of our loved ones.

But for me, the most powerful moment came on the reunion day at the farm, just before we sat down together for the evening meal. In the shadow of Grandpa's barn and on the farmstead he and Grandma had built, all 200 Arens family members attending bowed their heads in prayer to ask that familiar blessing on the food: "Bless us, O Lord, and these thy gifts, which we are about to receive, from thy bounty, through Christ, our Lord." Everyone recited the prayer in unison.

One of my cousins standing next to me looked up. "This is what a family farm is all about," he said in a low voice.

I nodded. And the family chimed in, "Amen."

Acknowledgments

I wish to thank my wife Donna, for her considerable support, love and patience throughout this project, her continuing encouragement along the way, and her unyielding insistence on proper use of the English language. I want to thank my children, Lauren, Taylor and Zachary, for their antics and inspiration. I thank my parents, Harold and Margaret, and my brother Paul and his wife, Jules, for their contributions, prayers and support.

I thank all of my neighbors and friends around Crofton, who have truly shown me what "love thy neighbor" really means, pitching in selflessly to help any time we've ever needed it. They are the best.

I thank my college buddies—Mike Roeber, Randy Micek, Ron Drozd, Anne Mohr, Bugs Rethwisch, Doug Moser, Keith and Audrey Berns, Brian and Christi Berns, Ron and Diane Kulwicki, Kurt and Lisa Jackson, Cindy and Marshall Johnson, and many others—who helped me deepen my spiritual base and have always lightened the load, not only in college but also in my life today.

I am so grateful to the gentle School Sisters of St. Francis and all of my teachers and mentors at St. Rose School and Crofton High School, as well as Kevin and Tweeter Henseler and my high school newspaper co-editor and pal, Duane Schieffer, for encouraging the writing bug in me at an early age.

I am blessed to live in a rural faith community in Crofton at St. Rose parish and school, where volunteerism, compassion and community are not just words but values put into practice each day.

Finally, I wish to thank my friends and mentors in the writing of this project, Mary Pat Hoag and Roger Holtzmann, for their honest observations, expertise and guidance.

Appendix

The Litany of St. Isidore

Lord, have mercy on us.
Christ, have mercy on us.
Lord, have mercy on us.
Christ, hear us.
Christ, graciously hear us.
God the Father of heaven, have mercy on us.
God the Son, Redeemer of the world, have mercy on us.
God, the Holy Spirit, have mercy on us.

Response to all below: pray for us.
Holy Mary…
St. Isidore…
St. Isidore, patron of farmers…
St. Isidore, illustrious tiller of the soil…
St. Isidore, model of laborers…
St. Isidore, devoted to duty…
St. Isidore, loaded down with the labors of the field…
St. Isidore, model of filial piety…
St. Isidore, support of family life…

St. Isidore, confessor of the faith...
St. Isidore, example of self-denial...
St. Isidore, assisted by angels...
St. Isidore, possessor of the gift of miracles...
St. Isidore, burning with lively faith...
St. Isidore, zealous in prayer...
St. Isidore, ardent lover of the Blessed Sacrament...
St. Isidore, lover of God's earth...
St. Isidore, lover of poverty...
St. Isidore, lover of others...
St. Isidore, most patient...
St. Isidore, most humble...
St. Isidore, most pure...
St. Isidore, most just...
St. Isidore, most obedient...
St. Isidore, most faithful...
St. Isidore, most grateful...

Response to all below: we beseech you, hear us.
Jesus, Our Lord...
That you would vouchsafe to protect all tillers of the soil...
That you would vouchsafe to bring to all a true knowledge of the
    stewardship of the land...
That you would vouchsafe to preserve and increase our fields
    and flocks...
That you would vouchsafe to give and preserve the fruits
    of the earth...
That you would vouchsafe to bless our fields...
That you would vouchsafe to preserve all rural pastors...
That you would vouchsafe to grant peace and harmony in our homes...
That you would vouchsafe to lift up our hearts to you...

Be merciful, spare us, O Lord.
Be merciful, graciously hear us, O Lord.

Response to all below: deliver us, O Lord.
From lightning and tempest...
From pestilence and floods...
From winds and drought...
From hail and storm...
From the scourge of insects...
From the spirit of selfishness...

Lamb of God, who takes away the sins of the world, spare us, O Lord.
Lamb of God, who takes away the sins of the world, graciously hear us,
   O Lord.
Lamb of God, who takes away the sins of the world, have mercy on us.
Christ, hear us.
Christ, graciously hear us.

Let us pray: Grant, O Lord, that through the intercession of Blessed
Isidore, the husbandman, we may follow his example of patience and
humility and so walk faithfully in his footsteps that in the evening of
life we may be able to present to you an abundant harvest of merit and
good works, you who lives and reigns, world without end. Amen.

A Prayer for Farm Families

by Curt Arens

Almighty God and Creator,

We thank you for your gift of life on the land.

We praise you for the good earth that we steward.

We honor you for your creation of wildlife and animals, plants and crops, fields and meadows, mountains and valleys, forests and plains, and for the elements of weather that we partner with each day.

We ask for the guidance of your Holy Spirit in helping us care gently for that earth you have created.

Help us be kind, selfless, generous, pious, loving, neighborly, honorable and dignified in your sight.

Help us follow the example of St. Isidore, the farmer of Madrid, and his loving farm wife, St. Maria.

Accept willingly our offerings of sacrifice, anxiety, uncertainty and fear.

Accept willingly our vows of love for one another as a family living on your land and caring for your creation.

Bless us, faithful Good Shepherd, and keep us always within the warm embrace of your love and kindness.

Amen.

Other Titles in the American Catholic Experience Series

The Spiritual Apprenticeship of a Curious Catholic
By Jerry Hurtubise
San Francisco attorney Jerry Hurtubise explores his childhood and young adulthood in Indiana, Illinois and California in a series of vignettes that are both touching and humorous. 96-page paperback, $9.95

Watching My Friend Die
The Honest Death of Bob Schwartz
by Mark Hare
Rochester, New York, journalist Mark Hare documents the lingering death of his friend Bob Schwartz—a high school teacher and songwriter—to pancreatic cancer. Winner of a 2006 Catholic Press Association Book Award in the "Family Life" category. 143-page paperback, $9.95

Living in Ordinary Time
The Letters of Agatha Rossetti Hessley
by MaryEllen O'Brien
Chicago theologian MaryEllen O'Brien uses the letters of Warren, Pennsylvania, laywoman Agatha Rosetti Hessley to explore the years immediately following the Second Vatican Council. 96-page paperback, $9.95

Finding My Way in a Grace-filled World
by William L. Droel
Rochester, New York, native and current Chicago resident Bill Droel, a leader of the National Center for the Laity, tells the story of his experiences of settling in a community of close-knit neighborhoods and parishes on the south side of Chicago. 112-page paperback, $9.95

A Light Will Rise in Darkness
Growing Up Black and Catholic in New Orleans
By Jo Anne Tardy
New Orleans native Jo Anne Tardy relates stories of growing up in the Al section of New Orleans.110-page paperback, $9.95